my revision notes

AQA A2
GEOGRAPHY

Michael Raw

With thanks to all the students whose valuable feedback helped develop this book.

Hodder Education, an Hachette UK company, 338 Euston Road, London NW1 3BH

Orders

Bookpoint Ltd, 130 Milton Park, Abingdon, Oxfordshire OX14 4SB

tel: 01235 827827

fax: 01235 400401

e-mail: education@bookpoint.co.uk

Lines are open 9.00 a.m.–5.00 p.m., Monday to Saturday, with a 24-hour message answering service. You can also order through the Hodder Education website: www.hoddereducation.co.uk

ISBN 978-1-4441-6296-7

First printed 2012
Impression number   5   4   3
Year   2017   2016 2015   2014   2013   2012

Cover photo reproduced by permission of ANK/Fotolia

Typeset by Dianne Shaw
Printed in India

Hachette UK's policy is to use papers that are natural, renewable and recyclable products and made from wood grown in sustainable forests. The logging and manufacturing processes are expected to conform to the environmental regulations of the country of origin.

P02057

Get the most from this book

Everyone has to decide his or her own revision strategy, but it is essential to review your work, learn it and test your understanding. These Revision Notes will help you to do that in a planned way, topic by topic. Use this book as the cornerstone of your revision and don't hesitate to write in it — personalise your notes and check your progress by ticking off each section as you revise.

☑ Tick to track your progress

Use the revision planner on pages 4 and 5 to plan your revision, topic by topic. Tick each box when you have:

- revised and understood a topic
- tested yourself
- practised the exam questions and gone online to check your answers and complete the quick quizzes

You can also keep track of your revision by ticking off each topic heading in the book. You may find it helpful to add your own notes as you work through each topic.

Features to help you succeed

Examiner's tips and summaries

Throughout the book there are tips from the examiner to help you boost your final grade.

Summaries provide advice on how to approach each topic in the exams, and suggest other things you might want to mention to gain those valuable extra marks.

Typical mistakes

The examiner identifies the typical mistakes candidates make and explains how you can avoid them.

Definitions and key words

Clear, concise definitions of essential key terms are provided on the page where they appear.

Key words from the specification are highlighted in bold for you throughout the book.

Exam practice

Practice exam questions are provided for each topic. Use them to consolidate your revision and practise your exam skills.

Now test yourself

These short, knowledge-based questions provide the first step in testing your learning. Answers are at the back of the book.

Check your understanding

Use these questions at the end of each section to make sure that you have understood every topic. Answers are at the back of the book.

Online

Go online to check your answers to the exam questions and try out the extra quick quizzes at **www.therevisionbutton.co.uk/myrevisionnotes**

My revision planner

Exam practice answers and quick quizzes at **www.therevisionbutton.co.uk/myrevisionnotes**

Unit 4

Option 4A

Option 4B

Exam practice and quick quizzes at **www.therevisionbutton/myrevisionnotes**

Countdown to my exams

6–8 weeks to go

- Start by looking at the specification — make sure you know exactly what material you need to revise and the style of the examination. Use the revision planner on pages 4 and 5 to familiarise yourself with the topics.

- Organise your notes, making sure you have covered everything on the specification. The revision planner will help you to group your notes into topics.

- Work out a realistic revision plan that will allow you time for relaxation. Set aside days and times for all the subjects that you need to study, and stick to your timetable.

- Set yourself sensible targets. Break your revision down into focused sessions of around 40 minutes, divided by breaks. These Revision Notes organise the basic facts into short, memorable sections to make revising easier.

Revised ☐

4–6 weeks to go

- Read through the relevant sections of this book and refer to the examiner's tips, examiner's summaries, typical mistakes and key terms. Tick off the topics as you feel confident about them. Highlight those topics you find difficult and look at them again in detail.

- Test your understanding of each topic by working through the 'Now test yourself' and 'Check your understanding' questions in the book. Look up the answers at the back of the book.

- Make a note of any problem areas as you revise, and ask your teacher to go over these in class.

- Look at past papers. They are one of the best ways to revise and practise your exam skills. Write or prepare planned answers to the exam practice questions provided in this book. Check your answers online and try out the extra quick quizzes at **www.therevisionbutton.co.uk/ myrevisionnotes**

- Try different revision methods. For example, you can make notes using mind maps, spider diagrams or flash cards.

- Track your progress using the revision planner and give yourself a reward when you have achieved your target.

Revised ☐

One week to go

- Try to fit in at least one more timed practice of an entire past paper and seek feedback from your teacher, comparing your work closely with the mark scheme.

- Check the revision planner to make sure you haven't missed out any topics. Brush up on any areas of difficulty by talking them over with a friend or getting help from your teacher.

- Attend any revision classes put on by your teacher. Remember, he or she is an expert at preparing people for examinations.

Revised ☐

The day before the examination

- Flick through these Revision Notes for useful reminders, for example the examiner's tips, examiner's summaries, typical mistakes and key terms.

- Check the time and place of your examination.

- Make sure you have everything you need — extra pens and pencils, tissues, a watch, bottled water, sweets.

- Allow some time to relax and have an early night to ensure you are fresh and alert for the examination.

Revised ☐

My exams

A2 Geography Unit 3

Date: ....

Time: ....

Location: ....

A2 Geography Unit 4

Date: ....

Time: ....

Location: ....

1 Plate tectonics and associated hazards

Plate movement

Earth structure

Revised

In cross-section the Earth comprises a series of concentric shells (Figure 1.1) of different chemistry, temperature and density. At the centre of the Earth, an iron-nickel **core** gives the planet its magnetic field. The core is divided into a liquid outer layer 2,200 km thick, and a solid inner layer 1,250 km thick. The **mantle**, which accounts for more than 80% of the volume of the Earth, surrounds the core. This layer consists of semi-solid rock (peridotite), 2,900 km deep. The upper part of the mantle, known as the **asthenosphere**, has plastic properties that allow it to flow under pressure.

The **crust** is the Earth's thin, rocky outer shell (Figure 1.2). There are two types of crust:

- **oceanic crust** — averaging 5 km in thickness and composed of dense basalt rock.

- **continental crust** — averaging 30 km in thickness (reaching a maximum 100 km beneath major mountain ranges). Its main constituent is granite, which is less dense than basalt.

The crust and the immediate underlying part of the mantle form a single unit called the **lithosphere**.

Pressure and temperature increase with depth in the Earth's interior, reaching over 3,000°C in the inner core.

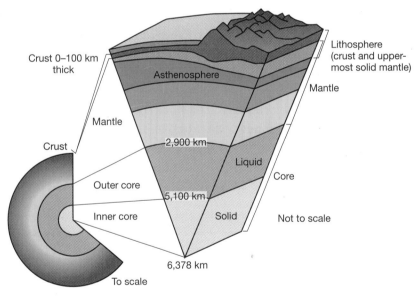

Figure 1.1 The Earth's structure

Labels: Crust 0–100 km thick; Mantle; Crust; Asthenosphere; 2,900 km; Liquid; 5,100 km; Outer core; Inner core; Solid; 6,378 km; To scale; Lithosphere (crust and upper-most solid mantle); Mantle; Core; Not to scale

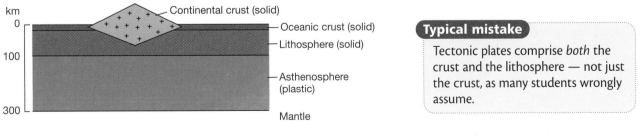

Figure 1.2 Cross-section through the Earth's crust, lithosphere and upper mantle

Labels: km; 0; 100; 300; Continental crust (solid); Oceanic crust (solid); Lithosphere (solid); Asthenosphere (plastic); Mantle

Typical mistake

Tectonic plates comprise *both* the crust and the lithosphere — not just the crust, as many students wrongly assume.

Plate tectonic theory

The Earth's crust and lithosphere are broken into seven large slabs and a dozen or so smaller ones known as lithospheric or **tectonic plates**. The global distribution of the main tectonic plates is shown in Figure 1.3. The boundaries between the plates are important zones of violent tectonic activity, including vulcanicity, seismicity, folding, faulting and mountain building.

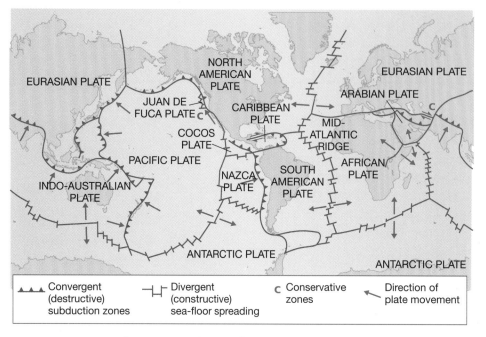

Figure 1.3 Distribution of main tectonic plates

Driven by convection currents in the Earth's interior, the tectonic plates slide across the plastic asthenosphere in a state of constant motion. New lithospheric crust is added by volcanic activity along mid-ocean ridges at **constructive plate margins**; old lithosphere and crust is transferred laterally away from these constructive margins (a process called **sea-floor spreading**) and is eventually destroyed in **subduction zones** or **destructive plate margins** (Figure 1.4). The result is a recycling of the oceanic crust over a timescale of approximately 200 million years. The outcome is not only tectonic activity along plate boundaries but also the slow procession of the continents across the Earth's surface, known as **continental drift**.

Typical mistake

It is wrong to assume that earthquakes and volcanic activity are confined to tectonic plate margins. Many earthquakes occur along fault lines thousands of kilometres from plate boundaries. Volcanic activity also occurs outside plate boundary zones, e.g. Hawaii, Canary Islands.

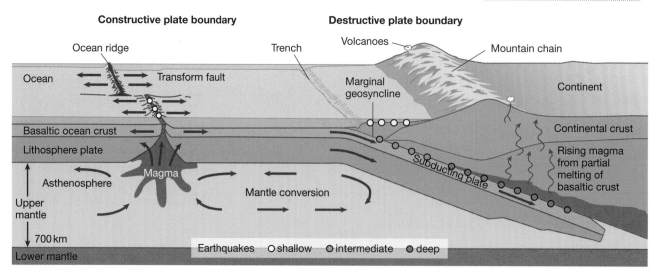

Figure 1.4 Constructive and destructive plate boundaries

Exam practice answers and quick quizzes at **www.therevisionbutton.co.uk/myrevisionnotes**

Evidence for plate tectonic theory

Continental drift is the most powerful evidence supporting the theory of plate tectonics. However, until the mid-1960s the evidence for continental drift was circumstantial. It included:

- the matching shapes of the continents on either side of the North and South Atlantic Ocean, which suggested that they might once have been joined
- ancient glacial deposits in the southern continents, formed during a past glacial period but now widely dispersed
- the same species of fossil plants and animals found in modern Africa and South America
- similar rock types and structures on opposite sides of the Atlantic in Brazil and west Africa, and northwest Europe and northeast North America

However, in the absence of a mechanism that could move entire continents, the scientific community remained sceptical about the idea of continental drift. The crucial evidence that led to the theory of plate tectonics came from studies of **palaeomagnetism** in the rocks of the ocean floor. Palaeomagnetism describes how iron particles in ancient lavas are 'frozen' in the direction of the Earth's magnetic field at the time they cooled and solidified. In the North Atlantic Ocean the polarity of basalt rocks on either side of the mid-Atlantic ridge showed a remarkable pattern. Because the Earth reverses its polarity at regular intervals (roughly every 400,000 years), symmetrical stripes (where iron particles are alternately aligned north then south) occurred in bands approximately 20–30 km wide. Moreover, the patterns on either side of the mid-ocean ridge formed a mirror image.

The implication was clear. New lithospheric crust, formed by volcanic activity and rising convection currents at mid-ocean ridges, slowly pushed the older crust sideways (at a rate of 1–2 cm per year). This process is called sea-floor spreading. As the ocean crust moves, the continents, like logs in an iceflow, 'ride' this natural conveyor. Thus sea-floor spreading drives continental drift.

> **Examiner's tip**
>
> Remember that the scientific evidence that clinched the theory of continental drift is sea-floor spreading.

Constructive plate boundaries
Revised

New lithospheric crust forms at constructive plate boundaries, where rising plumes of magma from the upper mantle stretch the crust and lithosphere (Figure 1.5). The result is intense volcanic activity, most often (though not always, as the example of Iceland shows) on the ocean floor. Volcanic eruptions build submarine volcanoes and submarine mountain ranges or **mid-ocean ridges**. Parallel faults associated with tension in the crust and volcanism produce **rift valleys** that extend for thousands of kilometres and separate the submarine mountain chains of mid-ocean ridges. Meanwhile, the mid-ocean ridges themselves are offset by huge lateral or **transverse faults**.

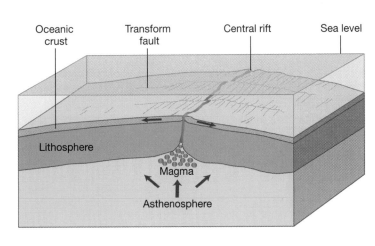

Figure 1.5 A constructive plate boundary

Destructive plate boundaries

Destructive plate boundaries or subduction zones are the sites where oceanic crust/lithosphere is destroyed (Figure 1.6). Subduction occurs when two tectonic plates converge. The older, denser plate is subducted. The subduction process involves:

● the descent of the subducted plate together with water and sea-floor sediments into the upper mantle

● the melting of the subducted slab and surrounding mantle rocks around 100 km or so below the surface

● the slow rise towards the surface of the melt (or magma), which is less dense than the surrounding rocks

● the eruption of lava, gases and ash at the surface through volcanoes and fissures

Where two oceanic plates converge (Figure 1.7), subduction forms an **island arc** such as the Kuril Islands in the north Pacific and the Lesser Antilles in the Caribbean. In contrast, the subduction of an oceanic plate at an oceanic–continental plate margin leads to the formation of **fold mountains** such as the Andes along the Pacific coast of South America (Figure 1.6). Destructive plate boundaries are also the location of volcanoes, earthquakes and ocean trenches.

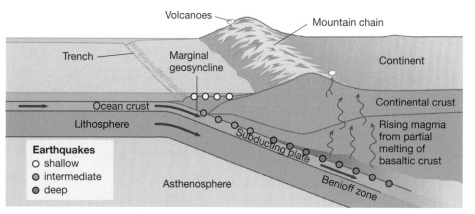

Figure 1.6 A destructive plate boundary

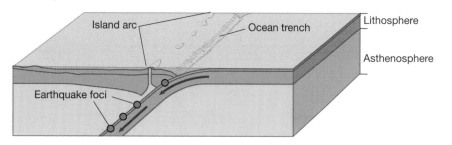

Figure 1.7 An oceanic–oceanic plate margin and island arc

Fold mountain ranges

The world's highest mountain ranges, including the Himalayas, Andes and Alps, are associated with plate convergence and the folding and uplift of sediments from the shallow ocean floor. The following sequence of events explains the formation of the Andes mountains:

● the Nasca oceanic plate converges with the continental South American plate

● sedimentary rocks, formed on the continental shelf and continental slope, are squeezed against the South American continent and crumple to form the Andean fold mountain range

- subduction of the Nazca plate produces huge **intrusions** of magma beneath the mountains, which create further uplift

The Himalayas were formed by the convergence of the Indo-Australian plate and the Eurasian continental plate. As the plates converged, the Tethys Sea narrowed until its sea-floor sediments were thrust 9 km above sea level into complex folds. This collision welded the continents of Asia and India together, producing a great thickness of continental crust. As a result volcanic activity is absent in the Himalayas.

Ocean trenches

Narrow trenches, hundreds of kilometres long and up to 11 km deep, occur on the ocean floor parallel to island arcs and fold mountain ranges. Ocean trenches mark the zone of subduction, where oceanic crust/lithosphere descends into the mantle. As it does so, the leading edge of the overriding plate is buckled to form a deep trench.

> **Examiner's tip**
>
> The landforms associated with plate boundaries depend on whether (a) a boundary is destructive/constructive/conservative, and (b) the nature of the lithosphere/crust on either side of the boundary, i.e. oceanic–oceanic, oceanic–continental, continental–continental.

Conservative plate boundaries
Revised

At **conservative plate boundaries**, two plates slide past each other with a shearing movement. This movement can be violent and may result in powerful earthquakes. However, volcanism is absent. In southern and central California, the boundary between the Pacific and North American plates is a conservative plate margin, known as the San Andreas fault. Earthquakes occur frequently along this fault line and present major hazards to metropolitan areas such as San Francisco and Los Angeles.

Hotspots
Revised

Hotspots are places where a plume of magma rises from the mantle, punches a hole through the lithosphere and crust, and erupts at the surface. They are usually associated with intense volcanic activity and eruptions of basaltic lava. Hawaii, the most volcanically active location in the world, lies on a hotspot at the centre of the Pacific plate. Other hotspots include Iceland and Yellowstone in Wyoming (USA).

> **Now test yourself**
>
> 1 List the main differences between oceanic and continental crust.
>
> 2 Draw a labelled diagram to show the structure of the Earth in cross-section.
>
> 3 Draw labelled diagrams to show the processes and landforms found at: oceanic–oceanic destructive plate margins; oceanic–continental destructive plate margins.
>
> 4 State three pieces of circumstantial evidence to support the theory of continental drift.
>
> 5 Describe the evidence for sea-floor spreading.
>
> Answers on p. 126

Vulcanicity

Destructive plate margins
Revised

Subduction occurs at destructive plate margins and results in volcanic activity. This is exemplified by the 'Pacific Ring of Fire'. Active subduction zones occur all around the margins of the Pacific plate, from New Zealand through South and North America and across the Bering Straits to Kamchatka, Japan, the Philippines and Indonesia (Figure 1.6). At depth, rising temperatures cause subducted oceanic crust, mixed with water and sea-floor sediments, to melt. The resulting magma then migrates slowly to the surface where it may erupt through volcanoes and fissures as viscous (thick) **andesitic lava** and **tephra** (ash/pumice). The high viscosity of andesitic magma traps steam and other gases, creating violent **explosive eruptions** (e.g. Mount St Helens eruption in 1980).

Constructive plate margins
Revised

Volcanic activity at constructive plate margins occurs as tension in the crust and lithosphere reduces pressure and allows magma to flow to the surface (Figure 1.5). Eventually lava, tephra and hot gases erupt through volcanoes and fissures.

Because constructive plate margins are found along mid-ocean ridges, most eruptions occur on the ocean floor. Volcanic activity at constructive margins also differs from destructive margins in a number of other ways

- lava is mainly **basalt** rather than andesite
- eruptions are quiet or **effusive** (rather than explosive) because basalt has low viscosity and allows gases to escape easily
- fluid basalt lava erupting on land (e.g. in Iceland) flows long distances before cooling and solidifying

> **Examiner's tip**
>
> The key to understanding contrasting eruptive behaviour at destructive and constructive plate margins is the viscosity of magma and the extent to which gases such as steam can escape easily.

Major extrusive activity — volcanoes
Revised

A volcano is an opening in the Earth's crust where molten rock and gases reach the surface. The ejecta or fragments thrown out by an eruption include lava, pumice, cinders, ash and gases. The nature of these materials is variable and explains differences in the shape of volcanoes and the nature of volcanic eruptions. There are two main types of volcano — Hawaiian-type or shield volcanoes and strato-volcanoes.

The classic volcano has a conical shape, as shown in Figure 1.8. The **cone** comprises layers of lava, ash and other ejecta erupted by the volcano. The **vent** occupies a collapsed hollow that, depending on its size, is known as a **crater** or **caldera**. Feeding the volcano, and located 3 or 4 km underground, is the **magma chamber**. Magma from the upper mantle fills this chamber before an eruption. The build-up of magma is detectable at the surface because the ground swells or inflates. Inflation can tear the crust apart to form rifts or fissures at the surface. Fissure eruptions are common in Iceland and Hawaii.

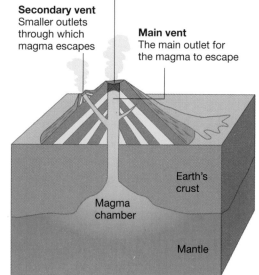

Crater
This is created after an eruption when the top is blown off the volcano

Secondary vent
Smaller outlets through which magma escapes

Main vent
The main outlet for the magma to escape

Earth's crust

Magma chamber

Mantle

Figure 1.8 A strato-volcano

Hawaiian-type or shield volcanoes

The Hawaiian Islands in the Pacific Ocean are one of the most active volcanic areas in the world. Located at the centre of the Pacific plate, volcanic activity is associated with a rising mantle plume or hotspot.

Today, the Hawaiian hotspot is centred over the Big Island (also called Hawaii). The Big Island has formed in the last 1 million years by eruptions from its five volcanoes. The largest — Mauna Kea and Mauna Loa — reach over 4,000 m above sea level and rise 9,000 m from the ocean floor. Kilauea is the most active volcano and has been in continuous eruption since 1983.

In profile the Big Island's volcanoes have the shape of a flattened dome. They are known as **shield volcanoes**. They are giants — at its base Mauna Loa is 120 km in diameter. However, its slopes never exceed a gentle 12°. Hawaii's shield volcanoes form because:

- most eruptions consist of lava rather than ash and gas
- the basalt lava has only a relatively small proportion of silica. It is therefore non-viscous, has a low gas content, and flows for long distances before cooling and solidifying
- eruptions are relatively gentle with little explosive activity

Strato-volcanoes

Strato-volcanoes consist of layers of ash and lava. They have steeper slopes than shield volcanoes and are more conical in shape. In contrast to shield volcanoes, the magma that forms strato-volcanoes — andesite — is viscous, with a high silica content. Magmas such as andesite create explosive products such as cinders and ash, and fewer lava flows. Also gases, such as steam, trapped by viscous magma cannot escape easily. As a result, pressures build up, which can cause eruptions powerful enough to blow a volcano apart. This happened at Mt St Helens in the northwest USA in May 1980 when a massive explosion destroyed the top 400 m of the volcano.

> **Examiner's tip**
>
> Shield volcanoes, formed from basalt lava, are also a feature of constructive plate margins (e.g. Iceland) as well as hotspots.

> **Examiner's tip**
>
> To achieve high levels, answers to exam questions on vulcanicity must show a good understanding and effective use of appropriate terminology. Key terms should be learned thoroughly.

Minor extrusive activity
Revised

Hydrothermal features such as geysers, hot springs and boiling mud form where magma and hot rocks close to the surface interact with groundwater. Yellowstone in Wyoming has half the world's hydrothermal features, including more than 300 geysers.

Geysers develop when groundwater held in porous rocks is superheated (i.e. to temperatures well above boiling point) by underlying hot rocks. As this superheated water moves towards the surface it dissolves silica from the surrounding rocks and lines rock crevices, creating a complex underground 'plumbing' system of pipes and reservoirs. Constrictions in this system cause superheated water to accumulate temporarily in underground reservoirs. As the superheated water moves towards the surface its pressure falls, and it eventually flashes into steam, and erupts at the surface as a geyser.

Hot springs form by similar processes but lack constrictions in their underground plumbing. As a result hot water flows continually and gently to the surface. **Boiling mudpots** develop where superheated water (which is in limited supply) combines with sulphur to form sulphuric acid. The acid then attacks surrounding rocks, reducing them to silica and fine clay.

> **Typical mistake**
>
> It is a common mistake to oversimplify the causes and nature of volcanic activity. Explanations must take account of (a) different processes operating at destructive and constructive plate margins, and at hotspots, (b) different products of volcanic eruptions — magma, lava, tephra, pyroclastic flows, gases, etc.

Major intrusive features
Revised

Batholiths are large-scale intrusive features. They were formed from huge masses of magma that cooled slowly within the Earth's crust. Subsequent erosion stripped away the overlying rocks, exposing part of the igneous mass at the surface. Because batholiths comprise resistant rocks such as granite and gabbro they form prominent uplands such as Dartmoor and the Cairngorms. **Bosses,** such as Shap in Cumbria, are similar to batholiths, but on a smaller scale.

Minor intrusive features

Dykes and **sills** are small-scale igneous intrusions that influence relief at a local scale (Figure 1.9). Dykes are thin sheets of igneous rock intruded at a high angle to the inclination of older surrounding rocks. Large numbers of dykes occur in the Inner Hebrides on the islands of Mull and Skye. Some dykes extend over long distances. The Cleveland dyke runs from Mull in western Scotland to within a few kilometres of the North Sea near Scarborough.

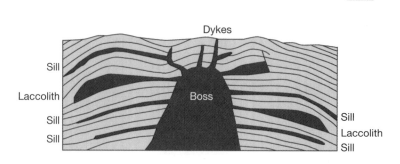

Figure 1.9 Minor igneous intrusions

Sills are thin horizontal sheets of magma. Where erosion of valley sides has exposed sills they often form cliffs and escarpments. The Great Whin Sill in northern England has a major effect on relief. It forms the steep cliffs at High Cup Nick in Cumbria; and a prominent escarpment followed by sections of Hadrian's Wall in Northumberland.

> ### Typical mistake
> Molten rock below the Earth's surface is known as magma. Magma extruded at the surface is lava. All intrusive igneous features have formed from magma that cooled below the surface.

Volcanic hazards

Natural events that cause death, injury and damage to property and infrastructure are known as **natural hazards**. Volcanic eruptions are examples of natural events that may become natural hazards. Large-scale natural hazards that result in major loss of life and widespread damage are called **natural disasters**.

The human impact of natural hazards can be explained by two concepts:

- **exposure** — the scale and frequency of the natural event and the number of people in the area affected
- **vulnerability** — the preparedness of a country or population to cope with a hazard

The variable impact of eruptions

Volcanic eruptions produce a range of hazards. The main types are shown in Figure 1.10. Their impact on people depends on three factors.

- The nature of the volcanic ejecta (i.e. lava, tephra, gas) and their violence. Gentle, effusive eruptions of lava, such as those found in Hawaii, pose little direct threat to human life, though lava flows may destroy farmland, buildings and infrastructure. On the other

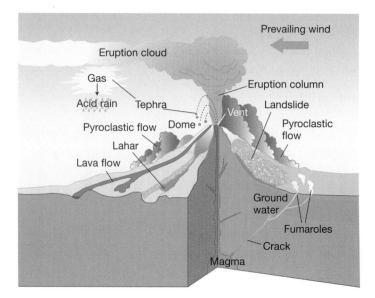

Figure 1.10 Volcanic hazards

hand, explosive eruptions of superheated gas and tephra often cause total devastation.

- The density of population and settlement in the vicinity of the volcano.
- Monitoring, warning systems and evacuation procedures for the population at risk. Compared with less economically developed countries (LEDCs), more economically developed countries (MEDCs) have sophisticated monitoring, early-warning and evacuation procedures. As a result, loss of life in developed countries is greatly reduced.

Lava flows

Although **lava flows** have occasionally been halted by spraying them with cold water (at Heimaey, Iceland in 1974 — see Table 1.1) and diverted by controlled explosions (e.g. Mt Etna, Sicily), most are unstoppable. Lava flows can cause enormous damage to property, although they are rarely a threat to human life. In 1990 a lava flow from Puu Oo crater in Hawaii buried the village of Kalapana. Further eruptions between 1992 and 1993 destroyed 181 homes, buried large areas of farmland and severed the main coastal road.

Pyroclastic flows

Pyroclastic flows are high-speed avalanches of hot ash, rock fragments and gas that destroy everything in their path (see the case study Soufrière Hills, Montserrat, 1995–2003). They can reach speeds of 200 km h^{-1} and temperatures of more than 1,000°C.

Lahars

Lahars are mixtures of water, rock, sand and mud that flow down valleys leading away from a volcano. They can be caused by:

- an eruption melting snow and icefields around a volcano's summit
- the rapid release of water following the breakout of a summit crater lake
- heavy rainfall washing away loose volcanic ash

Lahars are fast-moving and can travel long distances. They are particularly destructive because they follow valleys where settlements and population are often concentrated.

Jökulhlaups

Even more catastrophic than lahars are volcanic eruptions that occur beneath an icefield or glacier. Rapid melting of ice releases enormous volumes of water leading to massive floods. In Iceland these floods are known as **jökulhlaups**. Iceland's most recent major jökulhlaup occurred in 1996 following the eruption of the Grímsvötn volcano beneath the Vatnajökull icefield. A peak flow of 45,000 cumecs was recorded. The flood, which lasted for a week, destroyed several bridges and 10 km of the ring road that encircles the island. However, because the flood was expected, dykes were strengthened and people were evacuated so there was no loss of life or damage to settlements.

Mitigating volcanic hazards

Mitigation of volcanic hazards depends on monitoring and warning people of impending eruptions (Table 1.1). Monitoring includes recording seismic shocks, measuring ground inflation and collecting gas and lava samples. Hazard mapping can reveal areas most at risk from lava flows, lahars and pyroclastic flows. Preparedness is most advanced in MEDCs such as the USA and Japan, where it greatly reduces risks and loss of life.

Typical mistake

The distinction between natural events such as volcanic eruptions, and natural hazards, that cause death and damage to property, is often poorly understood. The result is often unfocused and irrelevant exam answers, and underachievement.

Examiner's tip

The starting point for an exam answer that assesses the impact of a volcanic eruption (or any other hazard) should be a recognition that impact is essentially the result of two factors — exposure and vulnerability.

Table 1.1 Mitigating volcanic hazards

Scheme	Description
Monitoring	Earthquakes and tremors develop as magma forces its way to the surface inside the volcano. These shocks are recorded by seismometers on the volcano. Gravity is also measured — as magma fills the reservoir beneath the volcano, gravity increases. Gases are sampled. Rising levels of sulphur dioxide and hydrogen chloride signal an impending eruption. Ground deformation (inflation) as magma accumulates within the volcano is further evidence of an imminent eruption.
Diversion of lava	Small lava flows have been successfully diverted away from centres of population. At Heimaey in Iceland, the fishing harbour was saved by spraying a lava flow with sea water.
Hazard mapping	The paths followed by ancient lahars and pyroclastic flows can be mapped from sediments.
Warning and evacuation	Lahar detection warning systems have been installed around Mt Rainier in Washington state. Detection triggers an automatic alert that initiates evacuation.

Case study Soufrière Hills, Montserrat, 1995–2003

Cause Montserrat is part of an island arc, formed by the subduction of the North American plate below the Caribbean plate. Montserrat owes its existence to the Soufrière Hills strato-volcano.

Volcanic hazards Pyroclastic flows, ash falls, debris avalanches and occasional lava flows. The volcano is explosive; its magma is thick, viscous and andesitic.

Impact Eruptive activity peaked in 1997 when 19 islanders were killed and Plymouth was destroyed by pyroclastic flows, ash falls and fires. Between 1990 and 2000 Montserrat's population fell from 11,000 to 6,500. Today, two-thirds of the island is uninhabitable and most fertile farmland in the south has been destroyed. Tourism, the former mainstay of the economy, has been ruined.

Exposure Soufrière Hills' eruptions are explosive and deadly pyroclastic flows carry high levels of risk. Given the small size of the island, Montserrat is densely populated. In 1990 nearly 11,000 people lived in areas at risk from pyroclastic flows and ash falls.

Vulnerability/management/responses Vulnerability was high because the eruptions occurred on a small island. However, this vulnerability was reduced by close monitoring of the volcano. Using data on seismic activity, volcanic gases and ground deformation, scientists have been able to issue early warnings and prepare people for evacuation. In the early eruptive stages, the area most at risk (i.e. the southern half of the island and the capital, Plymouth) was evacuated and designated an exclusion zone.

Case study Eyjafjallajökull, Iceland, 2010

Cause Volcanic activity caused by a constructive plate boundary (North American and Eurasian plates) and a spreading ridge across Iceland, coinciding with a hotspot. Volcanic activity over the past 3.5 million years has built Iceland.

Volcanic hazards Started as an effusive type of lava eruption from fissures. Later (mid-April) eruptions became more explosive because of an increase in dissolved gases in the magma moving to the surface, and steam formed when snow melted in contact with lava on the ice-covered summit caldera of Eyjafjallajökull. Melting snow caused flooding in adjacent lowlands (jökulhlaups) and ash falls. But the main hazard was a huge ash cloud up to 8 km high that drifted eastwards across Europe.

Impact In the vicinity of the volcano approximately 800 people were evacuated. In this area jökulhlaups damaged bridges, roads, farm buildings and farmland. There were no fatalities. The ash cloud closed Iceland's air space, resulting in a large temporary drop in the number of visitors, which hit tourism. The economic impact was felt in Europe and worldwide.

European air traffic was shut down for 6 days in mid-April, and intermittently until mid-May. The danger was ingestion of volcanic ash into jet engines which might result in catastrophic engine failure. Airport closure in Europe disrupted 10 million passengers at a cost of between £1.3 and £2.2 billion. Air freight was also badly disrupted.

Exposure Exposure to volcanic eruptions is high in Iceland, one of the most volcanically active regions in the world. However, low population densities and limited economic activity around Eyjafjallajökull reduced exposure. Exposure in mainland Europe to Icelandic eruptions is relatively low — the last major Icelandic eruption that affected Europe was the Katla volcano in 1787. Exposure also depends on wind direction, which determines the movement of the ash plume.

Vulnerability/management/responses In Iceland, levels of preparedness for volcanic eruptions are high. Eyjafjallajökull was monitored in the weeks before the eruption, and local people evacuated as minor earthquakes (suggesting an eruption was imminent) increased. The low frequency of the event in Europe meant that Europe was largely unprepared. Airports were summarily closed despite limited understanding of the effects of low- density volcanic ash on jet engines. The eruption prompted a review of procedures — whether closure of air space is necessary and whether aircraft can be re-routed rather than grounded — as well as further research into monitoring the movement of ash clouds and the effects of ash on jet engines.

Now test yourself

6 Describe two types of volcanic eruption.
7 Explain why some volcanic eruptions are violent and explosive.
8 Draw a labelled diagram to show the main features of a strato-volcano.
9 Describe four types of volcanic hazard.
10 State four ways in which volcanic hazards can be mitigated.

Answers on p. 126

Examiner's tip

A common examination theme is the management of natural hazards and its effectiveness. Good answers will make close reference to examples of specific hazardous events.

Seismicity

Causes of earthquakes
Revised

Earthquakes are vibrations (seismic waves) in the Earth's crust caused by the fracturing of rocks and sudden movements along fault lines. They result in violent shaking of the ground (the primary hazard), **liquefaction**, **landslides** and tsunamis (secondary hazards). The world's major earthquake zones correspond to plate boundaries. At these boundaries **inter-plate movements** cause tension, compression and shearing of the crust. Rocks under pressure eventually snap, resulting in crustal movements along fault lines that release huge amounts of energy as seismic waves. Major earthquakes can also occur thousands of kilometres away from plate boundaries. They are known as **intra-plate earthquakes**. Recent examples, caused by slippage along fault lines, include the Gujarat (2001) and Sichuan (2008) earthquakes.

The precise location of an earthquake within the crust is known as the **focus**. The point on the surface immediately above the focus is the **epicentre** (Figure 1.11). The destructive power of an earthquake is greatest close to the epicentre. Earthquakes of similar magnitude are generally more destructive when they occur near the surface (i.e. shallow quakes).

Seismic waves

Earthquakes produce two principal types of **seismic wave** — P-waves and S-waves. In the Earth's crust, P-waves travel at around 6–7 km s^{-1}, while S-waves travel more slowly (2.5–4 km s^{-1}). P-waves, like sound waves, consist of successive compression and stretching of particles in the rocks (Figure 1.12). The motion of these particles is parallel

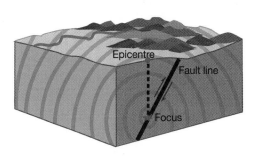

Figure 1.11 Earthquake focus, epicentre and fault line

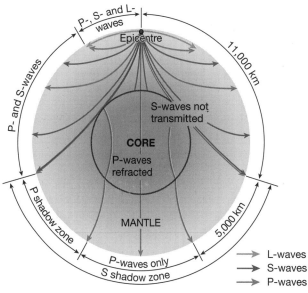

Figure 1.12 Seismic waves

to the direction of the wave. P-waves travel through both solids and liquids. S-waves are transverse waves, which means that particle motion is sideways. S-waves cannot travel through liquids.

Earthquake measurement

The **Richter scale** measures earthquake magnitude. Earthquakes range in magnitude from 2.5 to 9 on the Richter scale. Recent major earthquakes include Haiti (2010) magnitude 7, and northeast Honshu (2011) magnitude 9. **Seismographs** record the amplitude of earthquake waves, which radiate in all directions from the focus. These waves give a measure of the amount of energy released by an earthquake. The Richter scale is determined by the logarithm of the amplitude of seismic waves. In terms of energy release, a magnitude 7 quake is around 30 times more powerful than a magnitude 6 quake, and 900 times greater than a magnitude 5 event.

The **Mercalli scale** measures earthquake intensity, i.e. the impact of an earthquake on people and structures. The scale goes from 1 to 12, where 1 is instrumental (i.e. detected only by seismographs) and 12 is catastrophic, causing total destruction.

Typical mistake

It is a common mistake to assume that all earthquakes originate at plate boundaries. Many large quakes occur along fault lines at intra-plate locations, e.g. Sichuan quake in 2008.

Examiner's tip

A common exam theme is the relationship between earthquake magnitude and intensity. It is important to remember that earthquake magnitude is only one of several factors that influence the impact of earthquakes, and that overall, the relationship is weak.

Impact of earthquakes

Earthquakes damage buildings and infrastructure, and cause injuries and death. Large earthquakes can devastate an entire region and kill or injure tens of thousands of people. Collapsed buildings and other structures are the main cause of death and injury. In the aftermath of an earthquake, fire and disease may add significantly to the death toll.

More than one-third of the world's largest cities (most of them in LEDCs) are located in active seismic zones.

The risk to people from an earthquake (and other natural hazards) is summarised by:

$R = M \times P/V$

where: R = risk
M = size/scale of the earthquake
P = number of people living in the affected area
V = vulnerability (i.e. preparedness — building regulations, disaster planning, education, level of development)

Risk is therefore directly proportional to the size/scale of the event and the number of people living in the immediate area, and inversely proportional to society's preparedness for the hazard. Specific factors that influence earthquake impact include:

- time of day — an earthquake that strikes at night, when most people are asleep indoors, will cause more death and injury than a daytime quake
- population size and density — the larger and denser the population in an earthquake zone the more people at risk
- economic development — earthquakes are more damaging in poor countries, which lack the resources (a) to construct earthquake-proof buildings and infrastructure, and (b) to put in place effective emergency procedures to deal with the immediate impact of earthquake disasters

Examiner's tip

In any analysis of earthquake impact it is important that you support arguments with examples and case studies drawn from countries at contrasting levels of economic development.

Mitigating earthquake hazards

In contrast to other natural hazards, it is not possible to predict earthquakes and therefore provide populations with early warnings. And because earthquakes occur suddenly and unexpectedly, this makes them particularly deadly. However, we do know that in active earthquake zones such as California and Tokyo Bay, the longer the interval without an earthquake, the higher the probability of its occurrence and the greater its magnitude is likely to be. The main human response to earthquake hazards is to minimise (or mitigate) their impact.

Building design

Building technology is controllable and is a significant influence on the amount of damage, death and injury caused by earthquakes. In poor countries, few buildings are earthquake-proof. In rural areas in many LEDCs, traditional houses with heavy roof timbers and mud walls collapse easily, trapping their occupants. In urban areas, multistorey flats and reinforced concrete buildings — often built cheaply and with little regard for safety standards — may collapse, leading to high death tolls. Although many developing countries have strict building codes, the high death tolls in the Sichuan (2008) and Kashmir (2005) quakes show that these codes are often ignored.

Rich countries such as Japan and the USA, which straddle active earthquake zones, may avoid building high-rise structures in areas most at risk. However, in densely populated urban centres such as Tokyo or San Francisco this may not be an option. Instead strict building regulations are enforced to ensure that buildings and other structures survive the largest quakes intact.

Earthquake-proof high-rise buildings include designs with:
- steel frames and braces that twist and sway during an earthquake without collapsing
- foundations mounted on rubber shock absorbers
- deep foundations sunk into bedrock
- first-storey car parks allowing the upper floors to sink and cushion the impact
- concrete counter-weights on the top of buildings, which move in the opposite direction to the force of the quake

Disaster planning and prevention

In Japan, cities in earthquake zones have disaster plans to manage major earthquake events. The Tokyo Metropolitan Government is responsible for earthquake planning in the capital and aims to make the city 'disaster-resistant'. This involves upgrading millions of houses to make them fireproof; strengthening roads, expressways, bridges and public buildings; planning for evacuation to safe locations such as city parks (23 refuge sites in all); designating more than 3,000 public shelters to house 4.25 million people in an emergency; and educating people in disaster awareness and how to cooperate with other citizens to build a 'strong society' against earthquakes.

> **Examiner's tip**
> The construction of earthquake-resistant buildings has been a success in rich countries such as Japan and has saved countless lives. This success should be contrasted with the situation in many LEDCs, where building collapse during major earthquake events invariably leads to massive loss of life.

Case study Earthquake: Kashmir, Pakistan, 2005

Physical details
- Epicentre: 105 km north of Islamabad in the foothills of the Himalayas
- Magnitude: 7.6 Richter scale
- Depth: 26 km
- Date: 8 October 2005
- Local time: 8.50 a.m.

Cause Collision of Indian and Eurasian plates. Compression caused low-angled thrust faults. Movement along a thrust fault triggered the quake. Northern Pakistan and northern India are highly active seismic zones.

Hazards Primary hazard was violent shaking caused by seismic waves. Numerous landslides were secondary hazards. Some landslides blocked rivers and threatened flooding.

Exposure The earthquake was high magnitude and a major tectonic event. In a mountainous region with steep and unstable slopes, the quake was followed by massive landslides. Nearly 15 million people live in the affected area, where, over the past 50 years, rapid population growth has placed more people at risk.

Vulnerability 87% of the population is rural, many living in isolated villages in the Jhelum and Neelum valleys, with difficult access for emergency relief aid except by helicopter. Thousands were cut off for days and even weeks by landslides. People living above the winter snowline (400,000) were especially vulnerable. Poverty is widespread, with 30% living below the poverty line. The ability of poor people to help themselves and recover from the impact of a major quake is limited.

There was little preparedness and planning for disaster relief. The result was overcrowded refugee camps, poor sanitation, and a lack of heating, tents and clean drinking water.

Poor construction and engineering were blamed for most deaths. Few buildings were earthquake-proof. Most buildings (particularly in the countryside) were made of stone and cement blocks laid in weak sand or mud mortar. Others were dry stone or made from rounded cobbles sourced from local rivers. Many solid block concrete buildings in towns collapsed. Poor quality concrete and mortar, weak connections at corners and thin walls were also to blame.

Impact 87,000 people died and 3 million were made homeless. Most deaths were due to collapsed buildings. There were further deaths after the quake because of injury, exposure and disease. Many survivors were forced to sleep outside and cope with severe winter weather in the mountains. Half of all the buildings in Muzaffarabad were destroyed. There was extensive damage to roads, bridges, schools, hospitals and electricity infrastructure. Four-fifths of crops and half of all arable land were destroyed. 100,000 cattle were killed. Total cost of the quake: US$5 billion.

Case study Earthquake: Northridge, Los Angeles, 1994

Physical details
- Epicentre: Northridge in the San Fernando valley, north of Los Angeles in southern California
- Magnitude: 6.7 Richter scale
- Depth: 17.5 km
- Date: 17 January 1994
- Local time: 4.30 a.m.

The sediment-filled San Fernando valley and adjacent mountains comprise east–west trending structures.

Cause Movement occurred along a previously unknown (blind) thrust fault. Faulting, caused by convergence between the Big Bend of the San Andreas fault and the northwest motion of the Pacific plate, is responsible for numerous faults.

Hazards 10–20 seconds of strong shaking, which damaged buildings and infrastructure. The quake caused thousands of landslides and other slope failures.

Exposure The Los Angeles metropolitan area has a population of 16.5 million and an average density of 2,500 persons km^{-2}. The Los Angeles basin is one of the most seismically active in the USA. Exposure was therefore high.

Vulnerability California is the richest state in the world's richest country. Massive investment has helped to

reduce vulnerability despite high levels of exposure. The environment of southern California is designed for seismic resistance. There are stringent building codes, high levels of preparedness and emergency response procedures in place.

Impact The shallow focus of the quake and the densely populated, built-up Los Angeles basin account for the massive damage that resulted. The death toll was 57; 9,000 people were injured and 20,000 displaced from their homes. Most of the damage was concentrated in the San Fernando and Simi valleys. The economic cost of the quake was US$20 billion. Motorways collapsed at seven sites and 170 bridges sustained damage. The collapse of the I-5/SH-14 interchange near San Fernando was one of the costliest failures. Near the epicentre, well-engineered buildings survived the shaking without damage. Elsewhere there were many structural failures, which pointed to deficiencies in design and construction methods. Steel-framed buildings (including schools and hospitals) cracked and reinforced concrete columns were crushed. Several low-rise apartment buildings, constructed above open-air parking spaces, collapsed. Investigations following the quake showed a need to improve building codes.

Tsunamis
Revised

Tsunamis are huge waves at sea, usually caused by earthquakes. In the open ocean, tsunamis may be less than 1 m high and pass unnoticed.

However, as they approach shallow coastal waters, wave heights can increase dramatically (8–15 m) and overwhelm coastal settlements.

Tsunamis travel at speeds of up to 800 km h^{-1}. Early warning of tsunamis in coastal areas close to an earthquake epicentre may give little time for evacuation (e.g. northeast Honshu, 2011). However, if a tsunami forms thousands of kilometres away on the opposite side of an ocean, the authorities can give early warnings and evacuate areas at risk. Around the Pacific Ocean a tsunami warning system is operational. However, such a system did not exist in the Indian Ocean when the 2004 Boxing Day tsunami devastated the west coast of Sumatra (Indonesia) and parts of Thailand, causing 280,000 deaths. This tsunami was triggered by a 9.1 magnitude earthquake in the subduction zone off the west coast of Sumatra.

In addition to early warnings, other mitigating actions against tsunami hazards are:

- increasing public awareness (e.g. publication of maps showing susceptible areas, safety zones and direct routes to high ground; practising evacuation drills in high-risk areas)
- construction of sea walls, breakwaters and tsunami shelters
- land-use planning (restrict development in areas of greatest risk) supported by compulsory insurance

Case study   Tsunami: northeast Honshu, Japan, 2011

Physical details 4.46 p.m., 11 March 2011. Triggered by a magnitude 9, relatively shallow earthquake in the Pacific Ocean, whose epicentre was 130 km east of Sendai in NE Honshu. The resulting tsunamis were up to 6 m high.

Cause Movement on the fault line on the Pacific–North America plate boundary at the subduction zone marked by the Japan Trench. In some places the ocean floor moved up; in others it dropped down. Energy caused by the displacement of the water column created a series of giant waves. In shallow water near the coast, the waves steepened, with increasing amplitude and decreasing wavelength.

Hazards Extensive flooding of low-lying coastal areas up to 15 km inland. Giant waves and floating debris destroyed everything in their path.

Exposure Exposure to tsunamis is high. Four major tectonic plates meet in the Japan region, making it one of the most tectonically active places in the world. Earthquakes occur frequently and high-magnitude quakes are not uncommon. The coastal areas in NE Japan are low-lying and densely populated. Offshore fault lines, which can generate tsunamis, lie near to the coast.

Vulnerability Japan is a rich country. It has made extensive preparations to mitigate the effects of earthquakes and tsunamis. More than 40% of Japan's coastline is protected by concrete sea walls and breakwaters. Recently $1.5 billion was spent on a new sea wall at Kamaishi in NE Honshu, but this was overtopped by the tsunami, submerging the city centre. People are educated in earthquake and tsunami drills, and escape routes in the event of a tsunami warning are shown on street signs. Early-warning systems should give those at risk time to evacuate and escape to higher ground. However, because the quake occurred so near

to the coast people had little time to evacuate. In Iwate prefecture the tsunami had already arrived by the time the warning was issued. In Miyagi prefecture 6 m waves arrived just 10 minutes after the quake. Overall, despite impressive levels of preparedness, vulnerability is high.

Impact The economic cost of the tsunami disaster exceeded $100 trillion, making it the most costly natural disaster in history. For a number of weeks after the disaster the worst-affected areas experienced shortages of food, clean water and temporary accommodation.

By May 2011 the death toll was estimated at 30,000, with nearly 500,000 people made homeless. Damage to buildings, infrastructure and economic activity occurred on a massive scale. Entire towns were razed to the ground (e.g. Ayukawahama) and whole communities wiped out. Roads, bridges (17), railways, airports (Sendhai), harbours, sewage treatment works and water supplies were either badly damaged or completely destroyed. Financial markets were hit — the Nikkei index dropped by 1.7%, the Japanese government was forced to inject billions of dollars into the financial system to maintain market stability, and GDP fell by 0.5%.

Industrial production in Japanese companies such as Toyota, Honda, Sony, etc., was halted because of a shortage of parts. Overseas plants were also affected. Salt water contaminated huge areas of farmland and destroyed crops; the regional fishing fleet suffered severe damage. At Fukushima on the coast the cooling system in the nuclear power station failed, causing a near-meltdown in two reactors and releasing deadly radioactive material into the environment. Eventually the authorities established a 30 km exclusion zone around the plant. This nuclear disaster was comparable to the world's worst — at Chernobyl in 1986.

Now test yourself

11 What is the difference between an inter-plate and an intra-plate earthquake?

12 Construct a table listing and describing the factors that influence the human impact of earthquakes.

13 What are the primary and secondary hazards generated by earthquakes?

14 Review two possible responses for mitigating the impact of earthquakes.

15 Under what circumstances can major earthquakes also trigger tsunamis?

Answers on p. 127

Answers on p. 127

Examiner's tip

In any analysis of the threat posed by natural hazards, the exceptional nature of tsunamis — the limited effectiveness of any physical protection and their impact often thousands of kilometres from their origin — should be emphasised.

Check your understanding

1 Explain how the Earth's crust and lithosphere is in continuous flux, with a balance between formation and destruction.

2 What arguments would you use to convince a sceptic that continental drift is a reality?

3 Explain why volcanic activity is concentrated at and around tectonic plate boundaries.

4 Explain why inter-plate boundaries are often the site of major earthquakes.

5 How successful is mitigating action in reducing the human impact of volcanic, earthquake and tsunami hazards? Present your views in a table, with a series of bullet points.

Answers on p. 127

Exam practice

Section A

1 (a) Study Figure 1.4, which shows constructive and destructive plate boundaries. Describe and comment on the distribution of earthquakes in Figure 1.4. [7]

(b) Explain how the process of sea-floor spreading is responsible for continental drift. [8]

(c) With reference to two recent earthquakes, compare the management and responses to the events. [10]

2 (a) Study Figure 1.8, which shows a strato-volcano. Describe and comment on the main features of strato-volcanoes. [7]

(b) Explain why volcanic activity is concentrated on or close to tectonic plate boundaries. [8]

(c) With reference to two recent volcanic eruptions, compare their economic, social and environmental impacts. [10]

Section C

3 To what extent is the management of seismic hazards more successful in MEDCs than in LEDCs? [40]

4 'The impact of natural hazards owes more to human than to physical factors'. Discuss, with reference to either volcanic or seismic hazards. [40]

Answers and quick quiz 1 online

Online

Examiner's summary

✔ A thorough knowledge and understanding of the forms and processes associated with plate tectonics, vulcanicity and seismicity must be gained.

✔ Detailed case studies must be learned — at least two contrasting studies for volcanic eruptions and for earthquakes — and used to support descriptions, explanations and discussions.

✔ Sketch diagrams of, for example, types of plate boundary can be extremely useful aids to explanation.

✔ When confronted with general statements about the impact of natural hazards such as volcanic eruptions, earthquakes and tsunamis, the contrasting impacts on rich and poor countries should always be considered.

✔ Explaining the economic, social and environmental impact of natural hazards must take account of a range physical and human factors. Good exam answers avoid simplistic arguments and show an understanding of the complexity of the real world.

✔ The topic of plate tectonics and associated hazards is replete with technical terms. Knowledge of appropriate terminology and their accurate use are characteristic of higher-grade answers to exam questions.

2 Weather and climate and associated hazards

Major climate controls

The structure of the atmosphere Revised

The Earth's atmosphere is the thin envelope of gases which surrounds the planet (Figure 2.1). It extends for more than 100 km above the surface, though its density above 30 km is low. In cross-section the atmosphere comprises four layers defined by the behaviour of temperature with height. The lowest layer or **troposphere** is the most important. It accounts for three-quarters of the atmosphere's mass, and is where nearly all weather, including clouds, precipitation and winds, happens. The troposphere is defined by a decline or **lapse rate** of temperature with height which averages 6.5°C km^{-1}.

The gaseous composition of the atmosphere varies little with height. Nitrogen and oxygen account for 99% of the atmosphere by volume (Table 2.1). But **water vapour** and carbon dioxide, despite being present in only small quantities, have a huge influence on weather and climate. Water vapour is particularly important. It contributes up to 4% by volume near the surface but is largely absent above 10 km.

Table 2.1 Average composition of dry atmosphere by volume below 25 km (%)

Nitrogen (N_2)	78.08
Oxygen (O_2)	20.94
Carbon dioxide (CO_2)	0.03 (variable)
Ozone (O_3)	0.000006

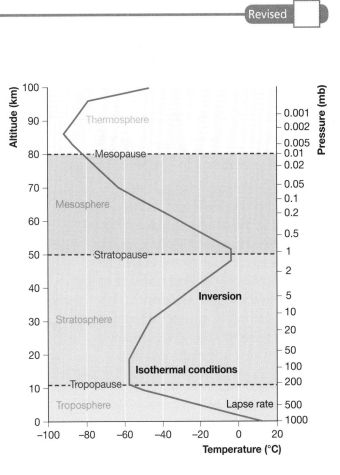

Figure 2.1 The structure of the atmosphere

The atmospheric heat budget Revised

At the global scale the Earth and the atmosphere form an energy system dominated by inputs and outputs of heat energy.

Energy inputs

Insolation (i.e. **in**coming **sol**ar radi**ation**) drives the global energy system. The sun is the source of this energy, which consists of electro-magnetic waves. This radiant energy is mainly short-wave and concentrated in the visible part of the spectrum. The atmosphere is largely transparent to short-wave insolation with only a small fraction being absorbed by gases in the atmosphere.

However, less than half of all insolation reaches the Earth's surface and is converted to heat energy. The main loss is due to **reflection** by clouds and is bounced back into space without being converted into heat. Losses also occur at the Earth's surface especially from highly reflective surfaces such as snow and ice. Other losses are due to **scattering** and **absorption** by clouds, dust and gas molecules such as water vapour and ozone in the atmosphere.

Energy outputs

Unlike the sun, the Earth is a cool body which emits long-wave (or infra-red) radiation. Only a small fraction (5%) of this terrestrial radiation escapes to space. The rest is absorbed by carbon dioxide, methane, water vapour and other **greenhouse gases** in the atmosphere. Most of this heat trapped in the atmosphere is eventually re-radiated to the Earth's surface.

Over a year and for the Earth as a whole, inputs of **solar radiation** are equal to outputs of **terrestrial radiation**. As a result the global energy system maintains long-term balance and stability.

Regional variations in the heat budget

While the heat budget maintains an equilibrium at the global scale, at the regional scale this balance disappears (Figure 2.2). Between the equator and latitude 40°, inputs of solar radiation exceed outputs of terrestrial radiation. This creates an annual surplus of heat energy in the tropics and subtropics. In middle and high latitudes the position is reversed: energy losses exceed gains, giving rise to an annual deficit.

These energy imbalances cannot be maintained: if they were then the topics and subtropics would get warmer every year, while middle and high latitudes would get colder. Surplus heat energy in low latitudes is transferred polewards by:

- planetary winds and the **general atmospheric circulation**
- warm and cold ocean currents — the **thermohaline circulation**

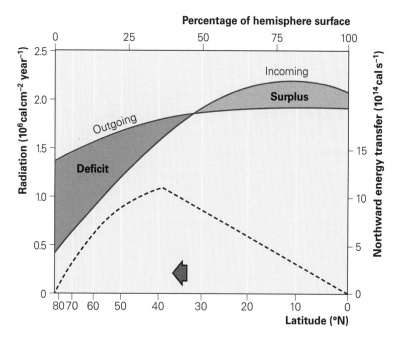

Figure 2.2 The global energy budget

The general atmospheric circulation

Geographical imbalances in the global heat budget, with an energy surplus in low latitudes and a deficit in middle and high latitudes, drive the general atmospheric circulation.

The tropics and subtropics

Between the equator and the subtropics the general circulation is dominated by two huge convective cells (known as Hadley cells), one in each hemisphere (Figure 2.3). Around the equator extreme **instability**

Instability is a thermal atmospheric condition where a parcel of air that is warmer than its surroundings rises freely through the atmosphere. The opposite condition is known as stability.

Exam practice answers and quick quizzes at **www.therevisionbutton.co.uk/myrevisionnotes**

causes warm air to rise at the **intertropical convergence zone** (ITCZ). On reaching the level of the **tropopause** the air diverges and flows towards the poles. As it does so it cools and its density increases until between latitudes 20° and 30° it slowly sinks towards the surface. Subsidence compresses and warms the air, preventing cloud formation. The result is clear skies, intense heating at the surface, permanently dry weather and the formation of the world's great tropical deserts and the subtropical 'high' pressure belt. Meanwhile the subtropical 'high' drives a return flow of surface air — the **trade winds** — back towards the equator, thus completing the convective cycle.

> The **tropopause** is the boundary between the troposphere and the stratosphere.

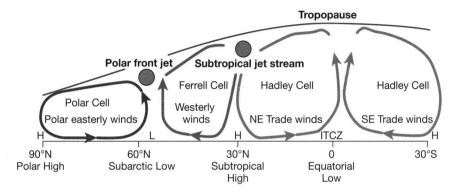

Figure 2.3 The general atmospheric circulation in low latitudes

Middle latitudes

The general circulation in middle latitudes is known as the **Ferrel Cell**. The weather is dominated by a westerly airflow in which migrating **depressions** and **anticyclones** are embedded. This westerly flow is linked to a narrow, fast-moving belt of air known as the **polar front jet stream** which encircles the Northern and Southern Hemispheres in a series of waves. The jet stream steers the air masses, depressions and anticyclones that are responsible for the unsettled weather of mid-latitudes.

High latitudes

The **Polar Cell** describes the atmospheric circulation in high latitudes. It is similar in structure to the circulation in the tropics and subtropics and is driven by radiative cooling, with air sinking near the poles and rising near latitude 60°. Cold dense air at the poles creates an area of permanent high pressure, producing an outward flow of surface easterly winds.

> **Examiner's tip**
>
> Think of the atmospheric circulation of planetary winds as a part of a system for evening-out imbalances in the heat budget between low latitudes (surplus) and high latitudes (deficit).

Planetary surface winds

Revised

Winds are horizontal movements of air that blow from areas of high pressure to areas of low pressure. At the global scale there are three planetary surface wind systems (Figure 2.3):

- the northeast and southeast trade winds, which blow equatorwards from the subtropical 'high' to the ITCZ
- the westerly winds, which blow polewards from the subtropical 'high' to the sub-Arctic 'low'
- the polar easterly winds, blowing outwards from high pressure over the poles

Surface winds exert an important control on climate. In middle and high latitudes, westerly winds bring characteristically unsettled weather, with 'families' of **depressions** alternating with ridges of high pressure or **anticyclones**. In the tropics and subtropics, the influence of the trade winds depends primarily on whether they blow onshore or offshore. Onshore trade winds in coastal Brazil and southeast Africa create humid conditions and generate heavy rainfall. In contrast, offshore winds in northeast and southwest Africa are dry and are partly responsible for hot desert climates.

Latitude
Revised

Latitude controls the angle of the sun and the amount of incident solar radiation received at the surface. The intensity of solar radiation is greatest in low latitudes where the sun is overhead for part of the year. This results in the temperature gradient between the tropics and the poles. Anomalies, such as the equatorial regions receiving less insolation than the tropical deserts, are due to the extent of cloud cover.

Latitude also affects seasonality. Around the equator seasonal variations in the sun's angle in the sky are less than half those in the British Isles. This explains the:

- small seasonal differences in temperature in low latitudes, but pronounced summer and winter differences in middle and high latitudes
- small seasonal variations in daylight hours in low latitudes and the large winter–summer variations in middle and high latitudes

> **Examiner's tip**
>
> Although latitudinal variation in solar radiation intensity is the main driver of climate, regional climates are controlled by a combination of other factors such as prevailing winds, ocean currents and altitude.

Oceanic circulation
Revised

Approximately 30% of the surplus heat energy in the tropics and subtropics is transferred polewards by warm surface ocean currents. Cold water eventually circulates back into lower latitudes via deep ocean currents.

Because of the warm North Atlantic Drift, western Europe has a large positive **temperature anomaly** in winter. This contrasts with eastern Canada on the opposite side of the Atlantic. Influenced by the cold Labrador Current, eastern Canada experiences severe winters with freezing conditions for several months and the formation of sea ice. The effects of ocean currents in the North Atlantic are also amplified by the prevailing westerly winds. In western Europe these winds are onshore, and spread the warming influence of the ocean far inland. In eastern Canada, they are offshore, and flood the region with cold continental air throughout the winter.

> A **temperature anomaly** describes a significantly higher or lower average temperature than the global average for the latitude.

Altitude
Revised

Hills and mountains modify temperature and precipitation. Temperatures decline with altitude at an average rate of 6.5°C km^{-1}. Consequently, hills and mountains are cooler than surrounding lowlands. In middle and high latitudes the growing season for crops and other plants is shortened and **evapotranspiration** is reduced. Lower temperatures also increase the amount and frequency of **precipitation**. As air masses are lifted mechanically across hills and mountains (or initial uplift triggers their free ascent) they cool and water vapour condenses to form clouds and precipitation.

Hills and mountains can also influence the climate in adjacent lowlands. The leeward side of an upland barrier is often drier and warmer than the windward side. This is due to air warming as it descends from the mountains to lower levels, causing clouds

> **Typical mistake**
>
> Air masses cool as they are uplifted over hills and mountains because they expand as pressure falls with height (they are also warmed by compression on descent). This temperature lapse is *not* due to any exchange of heat with the surrounding atmosphere.

Exam practice answers and quick quizzes at **www.therevisionbutton.co.uk/myrevisionnotes**

to evaporate and reducing the probability of precipitation. The Mojave Desert in California, situated to the lee of the Sierra Nevada, is an example of this so-called **rain shadow** effect.

The climate of the British Isles

Basic climatic characteristics
Revised

According to Köppen's classification, the British Isles have a temperate rainy climate. Precipitation occurs all year round; at sea level, the warmest month is everywhere above 10°C and the coldest above freezing. The climate of the British Isles is controlled by:

- its relatively high latitude (between 50°N and 60°N)
- its extreme oceanic location on the northwestern edge of the Eurasian continent, exposed to prevailing onshore westerly winds
- the North Atlantic Drift — a warm surface ocean current

Temperature

Throughout the year, temperatures are strongly influenced by (a) the warmth of the surrounding ocean and seas, and (b) latitude. The highest average July temperatures (18°C) occur in southeast Britain, an area that is nearly 10° of latitude further south than northern Scotland, furthest from Atlantic influence, and closest to the continent. Average summer temperatures decline with latitude, reaching just 12°C in northern Scotland. In winter, the west is milder than the east because the ocean is warmer than the land at this time of year.

Average temperatures fall in northern and western uplands such as the Scottish Highlands and the Lake District. For example, the average annual temperature on the summit of Ben Nevis (1344 m) is just −0.5°C.

Precipitation

Annual precipitation exceeds annual evapotranspiration everywhere in the British Isles. Despite the absence of a dry season, there are marked regional differences in precipitation (Figure 2.4). Precipitation is highest in the north and west because:

- most rain-bearing air masses and frontal systems originate in the Atlantic. As they track eastwards they shed much of their moisture as rainfall; they also warm and therefore become less humid as they move inland. Bristol in western England has a mean annual precipitation of 869 mm; Norwich, 300 km northeast, has just 658 mm
- uplands dominate the relief of the north and west. Mechanical uplift of air masses crossing these uplands triggers cooling, condensation and precipitation. Mean annual precipitation along the west Cumbrian coast at sea level is a moderate 850–900 mm; but 20 km inland, the central Lake District mountains (900 m) receive on average 4,000 mm a year. Northern and western mountains also have a rain shadow effect, reducing precipitation amounts in several parts of eastern Britain (e.g. eastern Scotland, northeast England).

> **Examiner's tip**
>
> Explanations of the geographical distribution of precipitation can be enhanced by the use of diagrams to show, for example, the effect of relief on precipitation.

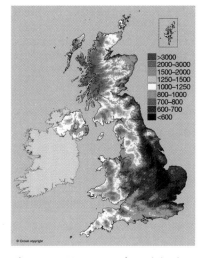

Figure 2.4 Mean annual precipitation in the British Isles 1971–2000

Wind

Like precipitation, wind and exposure increase with altitude. This is explained by reduced frictional resistance to airflow at altitude. Coastal areas also experience higher than average wind speeds. Again this is due to friction, with the relatively flat sea surface exerting less frictional resistance to the wind than the much rougher surface of the land.

Air masses affecting the British Isles — Revised

Air masses are large bodies of air covering thousands of square kilometres. They are defined by their uniform temperature, humidity and lapse rate.

The air masses affecting the British Isles originate from two **source regions**: subtropical and polar (Figure 2.5; Tables 2.1 and 2.2). Tropical air masses form in the subtropical 'high' in the mid-Atlantic (30–40°N) and in north Africa. Polar air masses develop within the Arctic Circle in northern Canada, northern Eurasia and Svalbard. Both subtropical and polar source regions are quiet, settled areas of permanent anticyclone where the air masses remain in prolonged contact with the ocean surface and the ground.

> **Examiner's tip**
>
> To predict the type of weather associated with an air mass you need to infer from its source region and track whether it will (a) be heated from below (making it unstable), (b) be cooled from below (making it stable), or (c) increase or decrease in humidity.

When an air mass leaves its source region its temperature, humidity and stability are modified. These modifications depend on the nature and direction of its **track**.

- Polar or arctic air moving towards the equator is heated from below, making it unstable (i.e. its lower layers become warmer than the air aloft).

- Tropical air moving polewards cools from below. Thus its surface layers are heavier and denser than the air above, preventing any vertical movement and keeping the air stable.

- Air masses crossing a sea surface evaporate moisture and become more humid. Meanwhile, air masses following continental tracks undergo little or no change in humidity.

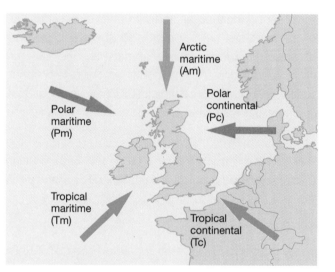

Figure 2.5 Air masses affecting the British Isles

Table 2.1 Classification of air masses and % frequency at Kew*

Source region	Track Maritime (m)	% frequency	Track Continental (c)	% frequency
Arctic (A)	Arctic maritime (Am)	6.5		
Polar (P)	Polar maritime (Pm)	34.7	Polar continental (Pc)	1.4
Tropical (T)	Tropical maritime (Tm)	9.5	Tropical continental (Tc)	4.7

* Does not include days when weather is dominated by depressions and anticyclones

Exam practice answers and quick quizzes at **www.therevisionbutton.co.uk/myrevisionnotes**

Table 2.2 Air masses and weather types affecting the British Isles

Air mass	Source region. Direction of approach	Stability	Winter weather	Summer weather
Arctic maritime (Am)	Arctic Ocean. Northerly	Unstable over the sea; becomes increasingly stable over land	Warmed by the ocean and unstable on reaching the British Isles. Northerly winds bring wintery showers to north, and north-facing coasts. Cold, with below-average temperatures and night-time frost. Cooling over the land often leads to clear skies and cold sunny conditions	Rare in summer. In April and May outbreaks of Am air give unseasonably low temperatures (e.g. maximum 9–12°C)
Polar maritime (Pm)	Northern Canada, Greenland. Westerly to northwesterly	Unstable over the sea and land	Warmed as it crosses the ocean. Unstable and humid. Rain showers most frequent in the west, often dying out inland. Blustery northwest wind. Above-average temperatures (c. 8–10°C max). Clear skies at night with frost and fog inland	Below-average temperatures (c. 15–18°C max) with a cool northwesterly airflow. Some rain showers in the west. Instability may trigger convectional activity inland, with rain, hail and thunder
Polar continental (Pc)	Eastern Europe, Siberia. Easterly	Unstable in winter; stable in summer	Very cold in its source region, but warmed as it tracks west. Warming due to contact with the North Sea leads to instability and wintery showers in eastern areas, especially near the coast. Inland, cooling causes instability to weaken; showers die out and skies clear. Low daytime maxima (0–3°C) with night-time frost	The airmass originates in a warm continental interior. The North Sea cools and stabilises the air, bringing advection fog and low stratus cloud to the North Sea coast. Inland the fog and cloud evaporate to give clear skies, unbroken sunshine and temperatures into the mid-20s
Tropical maritime (Tm)	Azores, Bermuda. Southwesterly	Stable	The airmass cools and becomes more humid as it tracks northeast across the ocean. Usually overcast (stratus) in western Britain with drizzle on higher ground. Temperatures are above average and may reach 15°C even in mid-January	Stratus often forms over the sea and there may be some advection fog particularly around southwest coasts. Inland the fog soon evaporates and most of the British Isles have clear skies with temperatures typically reaching 24–26°C
Tropical continental (Tc)	North Africa and Mediterranean. Southerly/ southeasterly	Stable in winter; unstable in summer	Rare in winter. Winter weather is typically cloudy (stratus) and unusually mild	The air mass is dry and hot and gives heatwave conditions (max >30°C). Low humidity with clear skies. Thunderstorms may develop if temperatures rise sufficiently. Visibility often poor because of dust and smoke particles blown in from the near continent

Depressions
Revised

Depressions are large, travelling low-pressure systems that dominate the weather in mid-latitudes (Figure 2.6). The main features of depressions are shown in Table 2.3.

Formation of depressions

A simplified sequence of events leading to the formation of depressions is as follows.

- Air moves from west to east at variable speeds in a wave-like pattern (**Rossby waves**) in the **polar front jet stream** — a narrow belt of fast-moving air that encircles the globe at the height of the tropopause in mid-latitudes.
- The jet stream accelerates as it flows around a trough and

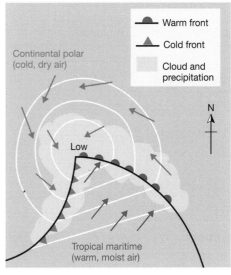

Figure 2.6 A depression model

decelerates around a ridge in the Rossby Waves — acceleration produces upper air divergence and low pressure at the surface.

- Low pressure at the surface sucks in warmer air from the south. This air rises through the lower atmosphere and forms the wedge of warm air or the **warm sector** in a depression.

Weather changes at the passage of a depression

Depressions bring windy, wet and changeable weather conditions. At **frontal zones** warm air is conveyed aloft, producing organised bands of thick cloud and prolonged spells of precipitation (Figure 2.7).

A typical sequence of weather changes associated with the passage of a depression is summarised in Table 2.3.

Ci = cirrus   Ns = nimbo stratus
Cs = cirro stratus   Sc = strato cumulus
Ac = alto cumulus Cu = cumulus
As = alto stratus   Cb = cumulonimbus

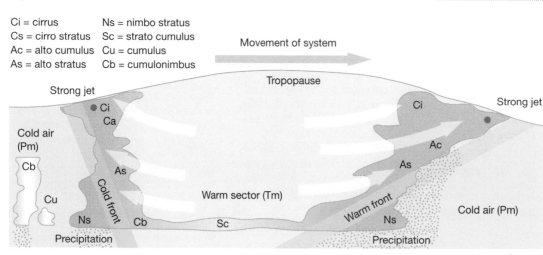

Figure 2.7 Cross-section through the warm sector of a depression

Table 2.3 Weather changes associated with the passage of a depression

	Approach of warm front	Passage of warm front	Behind warm front	Approach of cold front	Passage of cold front	Behind cold front
Pressure	Falls steadily	Stabilises	No change	No change	Rises abruptly	No change
Temperature	No change	Rises	No change	No change	Falls	No change
Cloud cover	Cloud thickens from cirrus, to alto-stratus and alto-cumulus, to nimbo-stratus and strato-cumulus	Thick cloud at low altitude	Cloud disperses; sky clears	Cloud thickens — nimbo-stratus and cumulo-nimbus	Thick cloud at low altitude	Sky clears abruptly. Individual cumuliform clouds develop in unstable Pm air
Precipitation	Begins slowly and continues steadily for several hours	Ceases	No precipitation	Heavy precipitation	Heavy precipitation. Chance of thunder	Intermittent showers in unstable airstream
Wind direction	No change	Veers (clockwise change)	No change	No change	Veers	No change

Anticyclones

Anticyclones are areas of high pressure. Their main characteristics are described in Table 2.4.

Exam practice answers and quick quizzes at **www.therevisionbutton.co.uk/myrevisionnotes**

Table 2.4 The contrasting features of anticyclones and depressions

Feature	Anticyclone	Depression
Surface pressure	High (1020–1050 mb), increasing towards the centre	Low (940–1000 mb), decreasing towards the centre
Wind direction*	Clockwise	Anticlockwise
Airflow	Diverges at surface; converges aloft	Converges at surface; diverges aloft
Vertical air motion	Subsides	Rises
Wind speed	Light winds that spiral outwards from the centre	Moderate to strong winds that spiral towards the centre
Precipitation	Generally dry; occasional drizzle	Wet — often heavy and prolonged
Cloudiness	Variable. Sometimes clear skies; sometimes overcast. Clouds have little depth (stratus)	Cloudy (nimbus, strato-cumulus). Deep clouds in bands along frontal zones
Air masses	A single air mass	Tropical and polar air masses separated by boundaries known as warm, cold and occluded fronts
Stability	Stable	Often unstable
Temperature gradient	Uniform	Strong contrasts across fronts
Speed of movement	Slow-moving	Mobile, fast-moving

*In the Northern Hemisphere

Source: Musk, L. (1988) *Weather Systems*, CUP

Formation of anticyclones

- Some anticyclones owe their development to cold, dense air aloft, which creates high pressure at the surface. 'Cold' anticyclones develop over Siberia and northern Canada during the winter.

- Warm anticyclones have a core of unusually warm air that extends to the upper troposphere (e.g. the subtropical Azores 'high'). Mobile warm anticyclones develop in mid-latitudes on the ridges and troughs of the polar front jet. Where the jet stream swings equatorwards, air speed slows, leading to convergence which forces air downwards towards the surface.

- Because the rate of convergence aloft exceeds the rate of divergence at the surface, high pressure develops. Mobile anticyclones often appear as ridges of high pressure between adjacent 'lows'.

Anticyclonic weather   Revised ☐

Anticyclones bring settled dry conditions, variable cloud and light winds. Once established they can persist for days or even weeks — a feature known as **blocking**.

Winter

In winter, subsiding air reaching the surface produces clear skies, daytime sunshine and night-time frost. Subsiding air that diverges *above* the surface often creates a **temperature inversion**. In these circumstances **stratus clouds** develop to give overcast conditions. If overcast conditions persist for several days they give rise to a weather phenomenon called **anticyclonic gloom**. In winter **radiation fog** often develops within the inversion layer at night. Because the sun is weak at that time of year, fog may remain all day.

A **temperature inversion** is an increase in temperature with height.

Stratus cloud is shallow, layer cloud formed by cooling of an air mass in contact with the surface. Clouds formed by convection, with flat bases and considerable vertical development, are known as cumulus clouds.

Summer

Clear skies under anticyclonic conditions in summer bring long hours of sunshine and temperatures above 20°C. Higher summer temperatures quickly disperse any inversion layer and evaporate overnight fog. Even so, anticyclones in summer are often cloudy, especially when onshore winds bring humid air from the Atlantic Ocean or North Sea.

Anticyclonic blocking

Anticyclonic blocking occurs when a large, slow-moving 'high' remains anchored over the continent and disrupts the normal westerly flow. Blocking forces mild, Atlantic air north or south of its normal track. As a result, the airflow becomes more northerly or southerly. Northerly flows introduce polar and arctic air, and below-average temperatures in all seasons. A southerly flow brings tropical air from North Africa and above-average temperatures. In summer this may give **heatwave** conditions. Blocking is often responsible for drought as well as unusually high or low sunshine amounts.

> **Typical mistake**
>
> It is mistake to think that anticyclones and high pressure are synonymous with fine weather (i.e. clear skies and sunshine). Although anticyclones usually bring dry weather, conditions are often cloudy and overcast.

Case study — Storm event: Cumbria floods, November 2009

Cause Prolonged stormy weather during a 7-week spell in autumn 2009. The jet stream remained stuck over the British Isles, bringing a series of deep depressions. Between 19 and 20 November a stationary cold front over Cumbria drew a conveyor of moist tropical air northwards from the Azores. Rainfall was intensified by the Lakeland Hills.

Rainfall and runoff November 2009 was the wettest November on record in the UK. The heaviest rain fell in the Lake District between 17 and 20 November. On 19 November 316 mm fell at Seathwaite in the central Lakeland — a record amount for a single day in the UK. The exceptional rainfall led to extreme flows on rivers draining mountainous catchments, such as the Derwent and Cocker. At Camerton in west Cumbria, the average 24 h flow for the River Derwent on 20 November was 561 cumecs — 20 times greater than the average.

Impact Widespread flooding occurred in Cumbria (2,200 properties affected). Floodwaters at Cockermouth reached a depth of 2.5 m, inundating more than 900 properties. Three bridges were completely destroyed and 20 others were closed temporarily. Several major roads became impassable because of flooding and landslips. Small retail and tourism businesses were badly affected. More than 100 farms were flooded, with loss of livestock. Extensive sheets of river-deposited gravels and silt covered fields. The port of Workington at the mouth of the River Derwent was closed because of river erosion and sedimentation of the harbour. The final bill for the Cumbria floods was more than £275 million. Damage to commercial and residential properties accounted for nearly 80% of the costs. The repair bill for damaged infrastructure was around £35 million.

Response Immediate response:
- During the floods 50 residents were rescued by the emergency services. Emergency relief centres were set up for hundreds of residents forced to evacuate their homes.

Response after 12 months:
- Cumbria's damaged infrastructure had been largely restored — 17 of the 20 bridges closed by the floods were open to traffic.
- The Environment Agency strengthened flood defences at Keswick and Cockermouth.
- Farmers received government financial assistance to remove sand, gravel and other debris left by the floods.
- The government's Flood Recovery Grant Scheme paid out more than £1 million to local businesses. Only 30 of Cockermouth's flooded businesses (226) remained closed.
- Most flood-damaged homes had been repaired and the majority of residents in Cockermouth's 691 flood-damaged homes had returned permanently.

Now test yourself

1. Explain why only 45% of insolation reaches the Earth's surface.
2. What is the main difference between solar radiation and terrestrial radiation?
3. Draw a diagram to show the main features of the Hadley Cell.
4. State two factors that influence the distribution of precipitation in the British Isles.
5. What is an air mass?
6. Name and describe one single factor that explains the differences in weather between depressions and anticyclones.

Answers on p. 128

Tropical savanna climate

Basic climatic characteristics
Revised

The tropical savanna climate is most extensive in Africa where it occupies a belt either side of the equator between latitudes 10° and 23°. It is found at similar latitudes in South America, India and Australia.

Compared with climates near the equator, the savanna shows greater thermal and hydrological seasonality. However, the average temperature of the coolest month is normally above 15°C. Depending on altitude, average monthly temperatures in summer vary from 25°C to 30°C (Table 2.5).

Seasons are differentiated more clearly by rainfall than by temperature. Typically the savanna climate has wet and dry seasons. The wet season approximates the period of high sun around the **summer solstice**. The period of low sun is one of prolonged drought.

Intertropical convergence zone (ITCZ)
Revised

The savanna's annual cycle of wet and dry seasons is explained by the movement of the ITCZ as it follows the overhead sun. Around the solstices the ITCZ is situated close to the tropics. Surface trade winds converge on the ITCZ from the **subtropical highs** in the Northern and Southern Hemispheres. These winds, together with intense surface heating create powerful updraughts at the ITCZ. Rising air cools, forming towering cumulo-nimbus clouds, thunderstorms and torrential downpours. The length of the wet season depends on latitude — the further from the equator, the shorter it is. After the solstice the ITCZ gradually shifts equatorwards, taking the rains with it, and the savanna comes under the influence of the subtropical high and prolonged drought.

> **Typical mistake**
>
> Seasons can be determined by rainfall patterns as well as temperature. In the tropical savanna there are two seasons: wet and dry.

Table 2.5 Tropical savanna climate: Gabarone (Botswana) 25°S, altitude 1,000 m

	J	F	M	A	M	J	J	A	S	O	N	D
Mean max °C	33	32	31	27	25	22	22	25	30	31	32	32
Mean min °C	19	19	17	13	8	4	4	7	12	16	18	19
Mean ppt (mm)	102	81	53	49	10	2	1	4	16	43	64	74

Tropical cyclones
Revised

Tropical cyclones are powerful storms that develop over warm oceans. In the Atlantic region they are known as **hurricanes**. Similar storms in east Asia and Australia are called **typhoons**, and in south Asia they are cyclones. Tropical cyclones are categorised on the Saffir-Simpson scale according to their power and destructiveness (Table 2.6).

Table 2.6 The Saffir-Simpson scale

Scale number	Central pressure (mb)	Wind speed (km h^{-1})	Storm surge (m)	Damage
1	> 980	119–153	1.2–1.5	Minimal
2	965–979	154–177	1.6–2.4	Moderate
3	945–964	178–209	2.5–3.7	Extensive
4	920–944	210–250	3.8–5.5	Extreme
5	< 920	> 250	> 5.5	Catastrophic

Occurrence

Tropical cyclones develop over oceans between latitudes 7° and 20°. Four conditions favour their development:

- plentiful water vapour
- light winds to allow vertical cloud development
- convergent winds at low level and divergent winds aloft
- sea surface temperatures (SSTs) of at least 26–27°C

These conditions occur in summer and early autumn in the tropical North Atlantic and North Pacific Oceans. As a result, the hurricane season in the Northern Hemisphere runs from June to late October. By November ocean waters are generally too cool to generate hurricanes.

Formation

The formation of tropical cyclones follows a sequence of events:

1 Clusters of thunderstorms develop over the ocean.

2 Some of these disturbances become better organised and develop an anticlockwise spin (in the Northern Hemisphere).

3 Surface pressure falls as the air, heated from below, becomes unstable and starts to rise.

4 Rising air cools, condenses and releases **latent heat**, warming the atmosphere and triggering further instability. This **feedback** intensifies the fall in pressure and causes surface air to be sucked towards the centre of the storm in powerful winds.

5 The central area of the cyclone behaves like a giant chimney: low pressure at the surface draws air inwards while high pressure aloft forces air outwards. In this way the cyclone gets a constant supply of fresh water vapour — the energy that drives the storm.

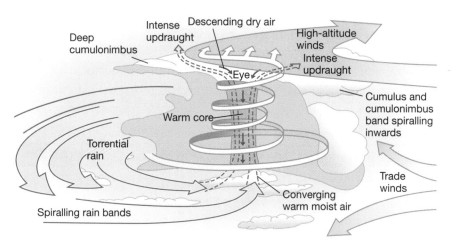

Figure 2.8 The anatomy of a hurricane

Decay

Tropical cyclones weaken and decay when they:

- move over cooler water, losing their supply of warm, moist air
- move over land, abruptly losing their power source (moist air)
- encounter strong **wind shear**

Impacts and hazards of tropical cyclones

Revised

Tropical cyclones are responsible for a number of related natural hazards, such as damaging high winds, heavy rain, river flooding and landslips, and **storm surges** and coastal flooding.

Storm surges up to 8 m high, caused by powerful winds and low pressure, temporarily raise sea levels and flood low-lying coasts.

Responses

Tropical cyclones are more closely monitored than any other natural hazard. Sophisticated measurements of temperature, humidity and wind speed, and tracking storm paths using satellites, aircraft, ships and buoys, allows accurate forecasts to be issued by agencies such as the US National Weather Service. The US National Hurricane Center measures and monitors hurricanes in the Atlantic and eastern Pacific.

Storm surges are the most deadly hazard caused by hurricanes. On the Gulf coast in the USA, major cities are protected by flood embankments or **levees**. In developing countries such as Bangladesh and India, storm shelters built on stilts provide temporary refuge from storm surges. Where storms are closely monitored, early warning may trigger mass evacuation, e.g. New Orleans in 2005 (Hurricane Katrina), Bangladesh in 2007 (Cyclone Sidr) and Houston in 2008 (Hurricane Ike).

Measuring and monitoring tropical cyclones

Monitoring begins in the early stages of storm development in the ocean using geostationary satellites, radar, ships and buoys moored at sea. Closer to land, direct measurements are made by aircraft and radiosondes. Data are fed into computer models which forecast storm intensities and tracks. National weather services issue warnings of approaching storms. In the USA two categories of warning are given: (1) *Hurricane Watch* —hurricane conditions are possible within the next 36 h, (2) *Hurricane Warning* —sustained winds of 119 km h^{-1} and above are expected within the next 24 h.

> **Examiner's tip**
>
> The vulnerability of countries to tropical cyclone hazards is closely related to differences in poverty and levels of development. This point should be emphasised in discussion on the variable impact of tropical cyclones and other natural hazards.

Case study Hurricane: Katrina, 2005

Background Hurricane Katrina hit the Gulf Coast of Louisiana on 29 August 2005. The hurricane intensified to category 5 as it moved west across the Gulf of Mexico, making landfall close to the city of New Orleans. It breached the levees separating New Orleans from Lake Pontchartrain, and caused disastrous flooding.

Exposure At its peak Katrina reached sustained wind speeds of 281 km h^{-1} with gusts exceeding 344 km h^{-1}. High winds and low pressure created a storm surge of 8 m at New Orleans. Exposure was increased by the 9.5 million people living in coastal counties along the Gulf of Mexico. New Orleans, surrounded by water (Lake Pontchartrain and the Mississippi River) and with large areas below sea level, is exposed to storm surges. Flooding was made worse by subsidence (caused by natural gas extraction) and the draining of wetlands in the Mississippi delta. By preventing annual floods on the river over many years, the levee system starved the delta of new sediment that would normally raise the level of the land surface.

Vulnerability and immediate response Katrina was monitored by the National Weather Service, which issued hurricane watches and warnings. Accurate predictions of the location and time of landfall were made. President Bush declared a state of emergency in Alabama and Mississippi before Katrina made landfall. The governor of Louisiana made a similar declaration on 26 August. Two days later New Orleans was evacuated. 350,000 people left the city. The city's superdome sports arena and conference centre provided refuges for the 150,000 people unable to flee. Although these measures did little to reduce the physical and economic damage caused by the storm, they saved thousands of lives.

The 560 km levee system, built to protect New Orleans from flooding, was not designed to withstand Katrina — a category 5 hurricane. Also, the levees were underfunded and had been poorly maintained for years before the disaster.

Impact and long-term response The storm killed 1,422 people, of whom 1,104 were in Louisiana. The storm surge that flooded New Orleans was responsible for most of the deaths. 80% of New Orleans was flooded. By 2009, 31% of all houses in New Orleans were still uninhabited and the city's total population was only 60% of that before Katrina. The damage (US$75 billion) made Katrina the costliest natural disaster in US history.

Following the Katrina disaster, a multibillion dollar scheme was funded to strengthen New Orleans' flood defences. The aim is to raise the levees and flood walls around the city to a level capable of withstanding a category 5 hurricane by 2011. New pumping stations are being built to remove water from the city. However, protecting areas of the city that lie up to 2 m below sea level is unsustainable in the long term. Rising sea level and the sinking delta may eventually mean the abandonment of these areas.

Case study Tropical cyclone: Nargis, 2008

Background Nargis was a category 4 tropical cyclone that struck southern Myanmar (Burma) on 2 May 2008.

Exposure Nargis generated several hazards, including sustained wind speeds reaching 210 km h^{-1}; a storm surge of 3.5 m; and torrential rain (600 mm). The cyclone hit the Irrawaddy delta. This delta lies close to sea level and is exposed to storm surges. It is also densely populated, supporting 3.5 million people.

Vulnerability Preparedness for tropical cyclone hazards is limited. This is partly due to poverty. The government provided no advance warning of the storm. There were no evacuation plans and no levees or storm shelters to protect against storm surges. The destruction of coastal mangrove forests increased vulnerability and risk to storm surges. Most people in the delta are peasant farmers, occupying flimsy houses. Having lost their crops and livestock to flooding, most had no resources to fall back on. Heavy dependence on rice farming and fishing added to their vulnerability.

Impact Globally, Nargis was one of the deadliest natural disasters of the 2000–2009 decade. The storm surge killed around 140,000 people, destroyed 450,000 homes and damaged 350,000 others. Three-quarters of health facilities and 4,000 schools were either ruined or badly damaged. In the worst affected areas, 85% of homes were destroyed. The cyclone also inflicted severe damage on the delta's rural economy and infrastructure. 600,000 ha of farmland were flooded, half of all livestock killed and food stocks, fishing boats and farm equipment destroyed. Large areas of farmland were contaminated by salt water. By mid-June there were serious food shortages and rice production in Myanmar fell by 6% in 2008. The estimated cost of the disaster was US$10 billion.

Response The government, unable to cope with the scale of the disaster, eventually allowed foreign aid agencies and workers access to the stricken areas. By July, 13 UN agencies and 23 NGOs were working in the delta. Priority was given to humanitarian aid to meet the immediate needs of survivors, i.e. food, shelter, clean water and medical supplies. Longer-term responses included rebuilding damaged roads, bridges, schools and hospitals, and restoring livelihoods. Because Myanmar is a poor country, recovery will depend on long-term support from the international community. Despite the massive international response, 12 months after the disaster 130,000 people were still living in temporary shelters and jobs remained scarce.

Urban climates

Energy exchanges   Revised

Towns and cities modify local energy exchanges to create microclimates that differ significantly from those in surrounding rural areas. Modifications to energy exchanges are due to:

- urban surfaces such as tarmac, brick and concrete, which absorb more short-wave radiation than natural surfaces such as vegetation and soil
- combustion from domestic heating, vehicles and factories, which add heat, CO_2, SO_2 and particulates to the atmosphere and reduce insolation amounts by 2–20%
- greater absorption of insolation in cities, which increases emissions of long-wave radiation by 5% or more at night, and 15–20% during daytime. The **pollution dome** over cities re-radiates much of this long-wave radiation back to the surface
- tall buildings and street 'canyons', which trap long-wave radiation
- lack of moisture in cities because of impermeable surfaces and rapid artificial drainage reducing the energy needed for evaporation and leaving more to heat the atmosphere
- an absence of strong winds, which would otherwise disperse the heat and bring in cooler air from the surrounding rural areas

Urban heat islands   Revised

Cities form 'islands' of warmer air, up to 5–10°C higher than the surrounding countryside (Figure 2.9). These temperature contrasts are greatest at night, especially with cloudless skies and light winds.

Exam practice answers and quick quizzes at **www.therevisionbutton.co.uk/myrevisionnotes**

Urban heat islands vary in shape and intensity. They are influenced by open spaces, rivers, building height and density, as well as the distribution of commerce and industry. Urban heat islands form a plateau of higher temperatures, which peak in the city centre. In some cities, a **temperature cliff** occurs at the edge of the urban area.

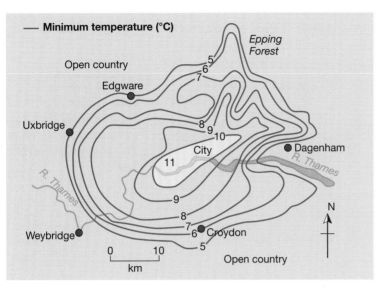

Figure 2.9 London's heat island in mid-May (light winds, clear skies)

> **Typical mistake**
>
> Remember that urban heat islands are not permanent features. They are most apparent at night and readily dissipate in windy conditions.

Precipitation
Revised

Rainfall

Cities can modify precipitation patterns. Convective precipitation requires a source of heat, water vapour and tiny particles known as **condensation nucleii**. Higher urban temperatures provide the trigger for uplift, cooling and condensation. At the same time particulates released by combustion act as condensation nucleii around which water droplets form. Because the convection process takes time, the effect of higher precipitation is often felt downwind, rather than in the city itself. On an annual basis, precipitation is on average enhanced 5–30% at downwind locations.

Fog

Average visibilities tend to be lower in city centres and improve towards the suburbs. But the frequency of thick fog (visibility less than 200 m) is greater in the suburbs and surrounding rural areas. This is due to (a) the heat island effect, which is most intense in the city centre, and (b) the greater concentration of condensation nucleii in cities, which results in the formation of smaller droplets that do not produce dense fog.

Air quality
Revised

Pollution

The urban atmosphere is polluted by gaseous and particulate emissions from heavy industry, electricity generation, domestic heating and motor vehicles. Pollutants responsible for poor air quality include sulphur dioxide, nitrogen oxide, ozone and suspended particulates (e.g. PM_{10}) such as soot. In China, air pollution is estimated to cause 400,000 premature deaths a year. It is worst in heavily industrialised coal-burning regions such as Shanxi province.

Air pollution is also a problem in most large cities in MEDCs. The main cause is vehicle exhaust emissions such as hydrocarbons and nitrogen oxide that react with sunlight to form **photochemical smog**. In the UK particulate pollution — mainly from traffic — causes 12,000–24,000 premature deaths a year. Pollution levels are highest during long spells of quiet, sunny weather. On 21 April 2011, London experienced its highest PM_{10} pollution levels since 2003. This was the 36th occasion since 1 January that pollution levels had exceeded safe limits. Under EU air quality laws the safe limits must not be exceeded more than 35 times in a calendar year. Unless London and other UK cities improve air quality, they face huge fines from the European Commission.

Pollution reduction policies

London, together with other large cities, has developed strategies to improve its air quality and comply with international legal limits. The main target is traffic. By 2015 London plans to:

- operate a low-emission zone across the capital and thus meet European standards for PM_{10} and nitrogen oxide pollutants
- place tighter controls on emissions from taxis and the city's own fleets of vehicles and office buildings
- promote less-polluting, hybrid and electric vehicles by providing preferential parking in central London and 25,000 electric charging points
- persuade people to use more sustainable transport, e.g. bicycles, by providing a bicycle hire scheme and bicycle lanes
- make idling of vehicles illegal

Winds

Revised

Urban areas modify the behaviour of the wind by: (a) increasing the roughness of the surface and the frictional resistance to air flow, and (b) street alignments that may channel or obstruct wind flow, depending on wind direction.

Large buildings act as obstacles to the wind. Air forced above and around buildings accelerates wind speeds. In the lee of tall buildings, turbulence often causes localised eddies and increases gustiness. At further distance downwind buildings have a sheltering effect and reduce wind speeds. Streets aligned at right angles to the wind experience turbulence and gustiness, while streets aligned in the direction of the wind channel the wind and cause an increase in average speed.

Examiner's tip

Your revision of urban climates must not only cover their causes and characteristics, but also their impact on human health.

Now test yourself

7 Describe the seasonal distribution of rainfall in the tropical savanna climate.

8 What is (a) the ITCZ, (b) the subtropical high?

9 Where and why do tropical cyclones form?

10 Name three natural hazards caused by tropical cyclones.

11 State three ways in which energy exchanges in urban areas differ from rural areas.

12 What is meant by the terms: (a) urban heat island, (b) photochemical smog?

Answers on p. 128

Global climate change

Evidence for climate change
Revised

20,000 years ago northern Europe was in the grip of an ice age (the Devensian). Ice sheets up to 1 km thick covered most of northern Britain and Ireland. Average annual temperatures were 5°C lower than today. Over the next 7,000 years or so, the climate gradually warmed. Ice sheets and glaciers melted and finally disappeared around 11,500 years ago.

Evidence for climate change comes from sea-floor sediments, ice cores, pollen analysis and dendrochronology (Table 2.7).

Table 2.7 Forensic evidence for climate change

Source	Description
Sea-floor sediments	The fossil shells of tiny sea creatures called foraminifera in sea-floor sediments are used to reconstruct past climates. The chemical composition of foraminifera shells indicates the ocean temperatures in which they formed.
Ice cores	Ice cores from the polar regions contain tiny air bubbles — recording the composition of the atmosphere in the past. Scientists can measure the relative frequency of hydrogen and oxygen atoms with stable isotopes. The colder the climate, the lower the frequency of these isotopes.
Pollen analysis	Pollen analysis allows reconstruction of past vegetational changes and palaeoclimates. Pollen diagrams show the frequency of pollen types (and therefore plant species) in the different sediment layers.
Dendrochronology	Dendrochronology is the dating of past events such as climate change through the study of tree ring (annule) growth. Annules vary in width each year depending on temperature and moisture availability.

Historical records also provide information about past climates. Europe experienced a so-called 'Little Ice Age' between the mid-fourteenth and early fifteenth centuries. The seventeenth-century diarist John Evelyn described frost fairs on the River Thames in London, events captured by contemporary artists such as Abraham Hondius. The sixteenth-century Dutch artist Pieter Breugel the Elder is famous for his winter landscapes of snow, frozen lakes and skaters. In 1783 scientists and naturalists in Europe and North America recorded the effects of the eruption of the Laki volcano in Iceland and noted the subsequent cooling in the summer of that year, and the exceptionally severe winter that followed.

Cause of global warming
Revised

There is conclusive evidence that the Earth's climate has warmed in the past 50 years (Figure 2.10). Globally, nine of the ten warmest years on record occurred between 2001 and 2010.

However, debate continues on whether **global warming** is a natural process or the result of human activities. The global climate has always undergone periodic change because of astronomical cycles, switches in surface ocean currents and volcanic eruptions.

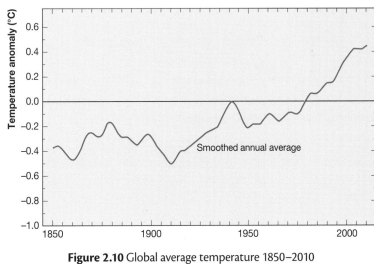

Figure 2.10 Global average temperature 1850–2010
Source: Met Office (based on Brohan et al. 2006)

Anthropogenic warming: the greenhouse effect

A strong correlation exists between the rise in average global temperatures and the amount of carbon dioxide in the atmosphere. Before 1800, average carbon dioxide concentrations were around 270 ppm. Today they average 390 ppm and are rising rapidly. This is due largely to the burning of fossil fuels, although deforestation and draining wetlands have also played a part.

The **greenhouse effect** explains the link between global temperatures and carbon dioxide levels. **Greenhouse gases** (GHGs) such as

Typical mistake

There is no doubt that the Earth's climate is warming; the debate is whether this warming is due to natural processes or human activities.

water vapour, carbon dioxide and methane, which occur naturally in the atmosphere, absorb and re-radiate around 95% of the Earth's long-wave radiation. However, large increases in carbon dioxide and other GHGs during the past 200 years have led to more absorption of long-wave radiation by the atmosphere. The result is an **enhanced greenhouse effect** responsible for global warming and climate change.

> **Examiner's tip**
>
> Any balanced discussion of the causes of global warming must make some reference to possible natural as well as human influences.

The global impact of climate change

Revised

Climate

The Intergovernmental Panel on Climate Change (IPCC) predicts an average rise in global temperatures of around 3°C by the end of the twenty-first century. However, continued growth in the use of fossil fuels could see global temperatures rising by as much as 5°C. Levels of warming will increase with latitude — temperatures in the Arctic and sub-Arctic will rise fastest, possibly by as much as 10–18°C.

Computer models predict major disruption to rainfall patterns in the twenty-first century. Although some regions will become wetter, large parts of North America, South America, southern Europe, Africa, the Middle East and central Asia will experience lower rainfall and more frequent droughts.

In mid-latitudes severe storms will be more frequent, increasing coastal erosion and flooding. Meanwhile, warmer conditions will intensify the water cycle, causing more evaporation, rainstorms and river floods. In the tropics and subtropics, warmer ocean waters will generate more powerful tropical cyclones.

Rising sea level

Melting glaciers and ice sheets, together with the **thermal expansion** of the oceans, are responsible for rising sea levels — levels rose by nearly 20 cm between 1900 and 2000. Current forecasts suggest an average rise in sea level of 40 cm by the end of the twenty-first century. This could spell disaster for countries such as Bangladesh, where 37% of the country is less than 3 m above sea level. Even worse, within the next 50 years island states such as the Maldives and Tuvalu could disappear altogether.

Sea level rise and stormier conditions also threaten coastal defences, especially in MEDCs. Future costs of maintaining sea walls and other defence structures could be prohibitive. Along some coastlines (e.g. eastern England) sea defences will be abandoned and nature allowed to take its course.

Water resources and farming

Some 16% of the world's population and one-quarter of global economic output could be hit by water shortages. With 98% of the world's glaciers currently retreating, areas that rely on meltwater (e.g. northern India) are most at risk.

> **Examiner's tip**
>
> Remember that the likely impact of global warming will be highly variable geographically. Economic status and physical conditions (altitude, climate, vegetation, fragility) will make some societies more vulnerable than others.

Any significant decline in rainfall in marginal farming areas (e.g. drylands in southern Europe and North Africa) could lead to land degradation and desertification. Other farming regions likely to be affected by drier conditions include commercially important areas such as the Prairies in the USA and Canada, and the Pampas in Argentina. As the climate dries, cereal production will slump. Some experts believe that production could drop by as much as 400 million tonnes a year, resulting in global food shortages.

Ecological impacts

Climate change puts huge pressure on natural **ecosystems**. Habitats will change and species will have to adapt by migrating either latitudinally or altitudinally. Thousands of species unable to adapt will face extinction.

Case study | The possible impact of global warming

On the British Isles

Water resources and farming Summer rainfall in southern Britain could decline by 40%. Temperatures could rise by 3°C by 2100. More frequent droughts could reduce crop yields. Transboundary pests and diseases affecting crops and trees will become more common in warmer, wetter conditions.

Population and infrastructure Flooding, especially in summer, could become more frequent, threatening homes, roads and power stations. More frequent heatwaves threaten higher mortality among the elderly.

Ecosystems Rising sea temperatures will force cold-water species such as cod and haddock further north. Southern species (e.g. tuna, sunfish) will become more common. Warmer conditions could push species endemic to northern Europe (e.g. ptarmigan, arctic hares, arctic-alpine plants) to extinction. They will be replaced by species from southern Europe (e.g. spoonbills, tongue orchids). Breeding of summer migrant birds could be disrupted by earlier springs.

Sea level Rising sea levels will increase erosion and coastal flooding in eastern and southern England. Sea defences will be strengthened along some coasts; elsewhere, reclaimed land will be abandoned to form mudflats and salt marsh.

On tropical savanna

Water resources and farming Average annual rainfall has been decreasing in Africa since 1970. Rainfall variability has also increased. Crop yields will decline further; pasture and forage will be reduced. Food insecurity will increase, with millions facing food shortages and malnutrition. Unirrigated cropland will be abandoned.

Population and infrastructure As marginal land becomes uncultivable, farmers will be forced to leave the land and migrate to overcrowded and impoverished towns and cities.

Ecosystems In regions such as the Sahel in sub-Saharan Africa, drought will lower plant productivity, and increase the risk of deforestation and overgrazing. The result could be widespread desertification. Feedback effects could disrupt the water cycle and cause further reductions in rainfall. Elsewhere, increased CO_2 levels may favour the growth of woodland and scrub at the expense of grass. Up to 40% of mammals in the savannas could face extinction by 2080 as a result of this ecological shift. Global warming could increase the risks of locust plagues.

Responding to climate change

Revised

Responses to climate change occur at international, national and individual scales.

International

Tackling global climate change requires international cooperation. So far, the only truly international initiative on global climate change is the **Kyoto Protocol** (1997). Under Kyoto some countries agreed to a 5.2% reduction in CO_2 emissions (based on 1990 levels) by 2012. Although an important first step, Kyoto has achieved only limited success. Only 37 of the 183 countries that ratified Kyoto have targets for a reduction in emissions and most signatories will fail to achieve their targets. Some of the world's largest producers of carbon dioxide (e.g. USA) did not sign up to Kyoto, while other major polluters such as China and India are exempt. So far no new international agreement to replace Kyoto has been reached.

Carbon trading offers an alternative approach. Under this scheme, businesses are allocated an annual quota for their carbon dioxide emissions. If they emit less than their quota they receive carbon credits, which can be traded on international markets. Companies that exceed their quota must either purchase additional credits or pay a financial penalty.

National

It is easier for individual governments to develop their own approaches to control carbon emissions. For example, governments may:

Examiner's tip

Global warming is a transnational problem which can only be solved by agreement among major countries. Vested national interests have so far made this difficult to achieve.

- give subsidies to promote renewable, carbon-free energy such as wind and solar power, as well as nuclear power
- impose a carbon tax on activities that use large amounts of fossil fuel
- encourage investment in new and/or cleaner technologies

Local

Individuals can contribute to lower carbon emissions by reducing their **carbon footprint**. Examples include energy conservation through better home insulation, walking or cycling to work, purchasing locally grown food, using recycling schemes and buying hybrid cars or more fuel-efficient vehicles. **Carbon offsets** encourage individuals (and companies) to take responsibility for their carbon emissions. Purchasing carbon offsets compensates for the emissions by funding an equivalent carbon dioxide saving elsewhere.

> ## Now test yourself
>
> 13 State three types of evidence to show that the global climate has changed in the past 20,000 years.
> 14 What is the greenhouse effect?
> 15 Why is global warming described as a transboundary issue?
> 16 List the possible environmental impacts of global warming.
> 17 What is: (a) the Kyoto Protocol, (b) carbon trading?
>
> Answers on p. 129

Check your understanding

1 Explain how the global heat budget drives the general atmospheric circulation.
2 How is the variability of the weather and climate of the British Isles influenced by air masses?
3 Draw a cross-section through the warm and cold sectors of a depression. Add labels to show the main features, and brief notes to describe the processes.
4 Describe how global warming and climate change could affect human activities and the environment in the tropical savannas.
5 Compile a table to summarise the natural hazards associated with tropical cyclones.
6 Explain how human activity gives rise to an enhanced greenhouse effect.

Answers on p. 129

Exam practice

Section A

1 (a) Study Figure 2.2 which shows the atmospheric heat budget. Describe how variations in the heat budget drive the general atmospheric circulation. [7]

  (b) Explain the role of global winds and ocean currents in the general atmospheric circulation. [8]

  (c) With reference to a recent example, discuss the impact of a major storm event in the British Isles. [10]

2 (a) Study Figure 2.5 which shows air masses that affect weather and climate in the British Isles. With reference to Figure 2.5, comment on the influence of air masses on the variability of weather and climate in the British Isles. [7]

  (b) Explain the origin and sequence of weather changes associated with the passage of a depression in the British Isles. [8]

  (c) With reference to case studies, evaluate the success of strategies to mitigate the impact of tropical revolving storms (i.e. hurricanes/cyclones/typhoons). [10]

Section C

3 Discuss the impact of tropical revolving storms and evaluate responses to them. [40]

4 Discuss the effectiveness of national and international efforts to tackle the problem of global climate change. [40]

Answers and quick quiz 2 online

Online

Examiner's summary

- ✔ Accurate understanding of the causes of weather and climate requires a thorough knowledge of physical processes.

- ✔ Sketch diagrams with labels and annotations are invaluable for explaining weather and climate phenomena such as the general atmospheric circulation, depressions, the effect of altitude on precipitation, etc.

- ✔ The correct use of terminology through learning keywords is particularly important in a topic such as weather and climate, which draws heavily on physical science.

- ✔ When revising, remember that the relationship between people and weather/climate, is an important focus of examination questions.

- ✔ High quality answers to extended examination questions on weather and climate will demonstrate knowledge of detailed examples and case studies.

- ✔ Answers to discursive questions on issues such as global warming require a balanced appraisal and a recognition that there are arguments on both sides.

3 Ecosystems: change and challenge

Nature of ecosystems

Structure of ecosystems

Revised

Ecosystems are communities of plants, animals and other organisms (the biotic component) and the environment (the abiotic component) in which they live and interact. These interrelationships bind the components of ecosystems into a coherent whole. We refer to this quality of wholeness as **holisticity**. The physical environment provides the energy, nutrients and living space that plants and animals need to survive.

Ecosystems are **open systems** (Figure 3.1). This means that both energy (e.g. solar radiation) and materials (e.g. minerals from weathered rocks, and water) cross ecosystem boundaries. Ecosystems exist at different scales, from a single oak tree to the planet itself. Those ecosystems that extend over large geographical areas (e.g. the tropical savannas) are known as **biomes**.

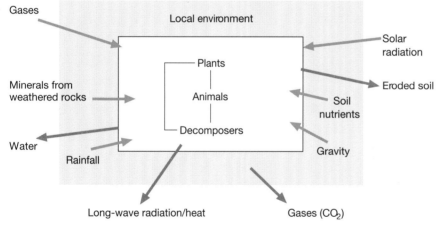

Figure 3.1 Inputs and outputs in an open ecosystem

> ### Examiner's tip
>
> The key to understanding the characteristics and behaviour of ecosystems is the relationship between living organisms and the physical environment.

Energy flows

Revised

Sunlight, captured by the leaves of green plants, is the primary energy source for most ecosystems. This energy is then transferred between organisms along simple **food chains**, or more often complex **food webs** (Figure 3.2). The flow of energy occurs in a number of stages or **trophic levels** (T).

- T1 — green plants intercept sunlight and, in the process of photosynthesis, convert sunlight, water, carbon dioxide and mineral nutrients into carbohydrates. Green plants (**autotrophs**) are the **primary producers** and ultimate source of energy in ecosystems.
- T2 — plant-eating animals or **herbivores** convert some of the energy from primary producers into animal tissue. Herbivores are the primary consumers in food chains/webs. **Consumers** are also known as **heterotrophs**.

- T3 — meat-eating animals or **carnivores** prey on herbivores. Carnivores, occupying the third tropic level in a food chain/web, are secondary consumers.

- T*n* — at the end of each food chain/web, there is a top or apex predator. Depending on the length of the chain, this animal may be a tertiary or quaternary consumer.

- At each trophic level, **detritivores** such as fungi and microbes decompose dead organic matter and animal faeces, consuming energy and releasing gases (carbon dioxide, methane) and mineral nutrients.

Energy is lost at each trophic level in a food chain/web. This is because organisms convert only a small fraction of the energy they consume into living tissue. Most energy is expended keeping the organism alive and is lost as heat in **respiration**. As a result:

- the number of trophic levels in food chains/webs is limited

- the **biomass** or dry weight of organisms declines at each trophic level (Figure 3.3)

- there is a reduction in the population of animals at each trophic level

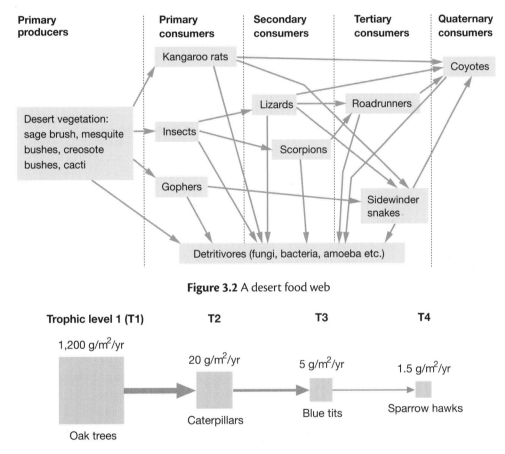

Figure 3.2 A desert food web

Figure 3.3 A simple food chain in an oak woodland

Nutrient cycles

Revised

Nutrients are the chemical elements and compounds needed by plants and animals. They cycle between the living organisms and the physical environment within ecosystems. There are three sources of nutrients.

- Rocks are the source of most nutrients. On **weathering**, rocks release nutrients such as potassium, calcium and sodium into the soil, which are absorbed by the roots of plants.

- Plants obtain some nutrients, such as nitrogen and carbon, directly from the atmosphere. Some mineral nutrients are also dissolved in rainwater.

- Eventually, most mineral nutrients return to the soil as **plant litter**. Dead leaves, roots and stems and other organic matter are decomposed and mineralised by fungi and microbes in the soil.

Nutrient cycles have variable speed and efficiency. In the tropical rainforest, high temperatures and humid conditions speed up decomposition and cause rapid nutrient cycling. This contrasts with the boreal coniferous forest, where low temperatures slow down decomposition. As a result, several years of leaf litter accumulate on the forest floor and nutrients, tied up in the litter, may rely on wildfires to release them.

> **Examiner's tip**
>
> A prominent feature of ecosystems is the interrelatedness of their biotic and abiotic parts and the number and complexity of these relationships. This quality should help you understand how ecosystems respond to change and why these responses are often difficult to predict.

Ecosystems in the British Isles

Ecological succession   Revised

The sequence of vegetation changes on a site through time is called **ecological succession** (Figure 3.4).

- **Primary succession** describes vegetation changes on sites previously uncolonised (e.g. sand dune, mudflat, bare rock). The first plants to colonise a site are known as **pioneer species**.
- **Secondary succession** describes vegetation changes on sites where the original vegetation cover has been destroyed (e.g. the fires in Yellowstone National Park, USA in 1988, which destroyed large areas of pine forest).

Ecological succession has the following characteristics:

- over time the physical environment, modified by plant growth, becomes increasingly attractive to a wider range of plant species
- there is a progressive increase in nutrient and energy flows
- biodiversity increases
- **net primary production** increases

> **Examiner's tip**
>
> Think of primary succession as starting from scratch (e.g. on a sand dune, a mudflat, a lava flow). Given the harsh environmental conditions at the outset, primary succession takes much longer than secondary succession.

> **Net primary production** is the amount of energy fixed in photosynthesis minus the energy lost in respiration. It is measured in grams per square metre ($g\ m^{-2}$) per year.

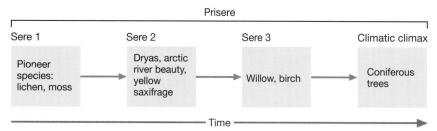

Figure 3.4 Primary succession in a recently deglaciated area

Climax vegetation   Revised

During ecological succession each vegetation stage or **sere** modifies the environment, allowing new species to invade, compete and dominate. Eventually the vegetation reaches a state of balance with the environment. This is known as the **climax vegetation**. Providing environmental conditions remain stable (e.g. climate remains constant), the climax vegetation will persist indefinitely.

The nature of climax vegetation is often controlled by a single environmental factor. When climate is the controlling factor, the vegetation is referred to as a **climatic climax**. Examples of climatic climax forms include biomes such as the tropical rainforest and the temperate deciduous forest. Where climax vegetation is most strongly influenced by a non-climatic factor such as slopes or soils, it is known as a **subclimax**. Climax vegetation that owes its main characteristics to human activities (e.g. heather moorland) is a **plagioclimax**.

Halosere

Revised

Vegetation plays a vital role in the development of salt marshes. Many marshes demonstrate a zonation of species that is closely related to height above sea level, as shown in Table 3.1. Such a plant succession in a saline, waterlogged environment is known as a **halosere**. It is best seen between the mean low water mark and the strand line on open marshes in estuaries.

Table 3.1 Plant zonation on salt marshes

Environment	Environmental conditions	Plants
Mudflats	High salinity levels; low oxygen levels in mud; high turbidity; long periods of inundation on each tidal cycle	No plants, only algae
Low marsh	Less hostile conditions than mudflats, but salinity and turbidity still high and oxygen levels low. Tidal inundation shorter	*Spartina* (cord grass) and *Salicornia* (glasswort) are the two common pioneer species. Sea blite (*Suaeda maritima*) and sea purslane (*Halimione portulacoides*) are found on better-drained areas (e.g. edges of creeks)
High marsh	Flooding only occurs on spring tides. Salinity levels are relatively low and soil develops	Wide variety of species, including salt marsh grass (*Puccinellia*), sea rush (*Juncus maritimus*), sea lavender (*Limonium*), sea aster (*Aster tripolium*), sea blite (*Suaeda maritima*) and sea purslane (*Halimione portulacoides*)

Primary ecological succession is responsible for the growth of salt marshes and follows a number of stages.

● Colonisation by pioneer species such as cord grass and glasswort. These plants slow the movement of tidal currents, which promotes rapid sedimentation (1–2 cm year^{-1}). Their roots help to stabilise the mud.

● Through accretion of sediment the marsh increases in height, reducing the period of inundation on each tidal cycle. Thus conditions become more favourable for the invasion of other, less-tolerant species. Biodiversity and plant cover increase. Plants such as sea rush, sea aster, sea lavender, salt marsh grass and common scurvy grass begin to dominate.

● The marsh height stabilises approximately 1 m above the mean high-tide mark. With only occasional inundation on the highest spring tides, vertical accretion ends, salinity levels drop, and soil starts to develop.

Temperate deciduous forest biome

Revised

The temperate deciduous forest once covered most of western Europe, eastern North America, eastern Asia, southern South America, southeast Australia and New Zealand. Little remains of the deciduous wildwood that once occupied most of lowland Britain. Over the past two millennia the forest has been cleared for farming and other economic activities.

Climate and soil

The climate of the temperate deciduous forest, strongly influenced by the ocean, is mild and humid. Average temperatures of the coldest months are above freezing, and in high summer average between 15° and 20°C. The result is a growing season that lasts seven to eight months. Precipitation occurs all year round and there is no dry season.

Most soils are **leached** and slightly acidic (pH 5.5–6.5), though soil acidity varies with **parent material**. Oak trees, which are deeply rooted, support a rich nutrient cycle and return a mild leaf litter to the soil each autumn.

Plants and animals

Broadleaved trees such as oak, birch and ash, which dominated the primary forest in the British Isles, form a canopy of 20–30 m. Below the canopy an understorey of shade-tolerant small trees such as elder, hawthorn, hazel and holly form a subdominant layer. Ground vegetation comprises a wide range of spring-flowering plants such as bluebells, dog's mercury, and celandine, as well as ferns and tree seedlings. Immediately above the soil surface there is a layer of moss and lichen. A rich ground flora flourishes in the early spring when the trees are leafless and strong sunlight floods the forest floor. The deciduous habit of trees is thought to be an adaptation to conserve moisture, which is less available in the winter months, and to the declining efficiency of water–gas exchanges in old leaves.

Insects and larvae, as well as small rodents such as shrews and voles, occupy the base of the food chain (Figure 3.5). Most large forest herbivores such as bison and wild boar are extinct in the British Isles. So too are the top mammalian predators — wolves, lynx and brown bears. There is a wide variety of bird life — seed-eating finches; insectivores such as woodpeckers and wrens; and raptors such as owls and sparrowhawks. **Biodiversity** is relatively high, though greater in North America than in Europe. Net primary production during the growing season is also high, though the overall average — 1,200 g m^{-2} per year — is around that of the tropical rainforest.

> **Leaching** is the removal from the soil of dissolved nutrients by rainwater. It is responsible for soil acidity.
>
> **Parent material** is rock or sediment that, through in situ weathering, provides the mineral fraction of the soil.

> **Examiner's tip**
>
> Few areas of climax deciduous forest survive in the British Isles. Most that do are protected as ancient woodlands. In discussion it is worth noting that natural vegetation has been altered, not just through forest clearance but also through the introduction of alien species.

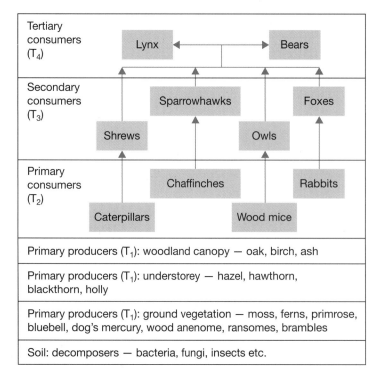

Figure 3.5 Structure of a temperate deciduous forest ecosystem

Exam practice answers and quick quizzes at **www.therevisionbutton.co.uk/myrevisionnotes**

Heather moorland

Revised

Ecosystem

Heather moorland is a plagioclimax formation, whose development and survival depends on human activities. A large proportion of the world's heather moorland is in the uplands of the British Isles where it covers nearly 5,000 km². Heather is the dominant moorland species. It consists of several low-growing, woody plants such as ling (*Calluna vulgaris*) and cross-leaved heath (*Erica tetralix*).

Heather moorland has limited biodiversity, short food webs (Figure 3.6) and low productivity. This is due to:

- severe environmental conditions in the uplands, e.g. high precipitation, low temperatures, poor drainage, acidic soils
- management that favours the growth of heather and a handful of animal species

Primary consumers include insects, insect larvae and red grouse. Ground-nesting birds such as golden plover, curlew and snipe, and small mammals such as pygmy shrews occupy the secondary consumer level. Foxes, stoats, hen harriers and merlins are at or near the top of the food web.

Management

Heather moorland is a managed ecosystem that owes its main characteristics to human activity. Management keeps the vegetation at an early stage of succession, to promote the growth of heather at the expense of competitors such as grasses, gorse and small trees (e.g. birch, rowan). The main purpose of management is to create a habitat for red grouse to support commercial shooting.

Red grouse, a subspecies unique to the British Isles, are completely dependent on heather for food and shelter. Management involves controlled burning of heather on a 15 year rotation. This has a number of effects:

- it stimulates the growth of young heather plants, which have the greatest nutritional value
- it encourages seed germination
- it keeps competing species in check

Heather moorland is also managed by controlling predation of the grouse by foxes, stoats, and crows, and even the illegal killing of birds of prey. Light grazing by sheep is also encouraged to prune the heather.

> **Typical mistake**
>
> It is widely assumed that heather moorland is a natural ecosystem. Evidence to disprove this is the replacement of heather with grasses, trees and scrub when management is removed.

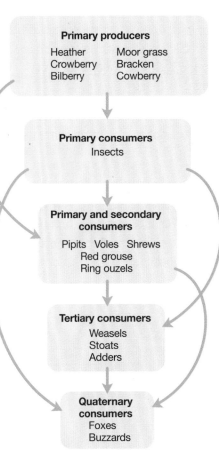

Figure 3.6 Heather moorland food web

Now test yourself

1 In what sense are ecosystems 'holistic'?
2 Draw a simple food chain to show energy transfer in a temperate deciduous woodland.
3 Name the principal nutrient stores in ecosystems.
4 What is the difference between primary and secondary succession?
5 Define the following: climatic climax, subclimax and plagioclimax vegetation.
6 What problems do pioneering plant species face in colonising tidal mudflats?
7 State four ways in which heather moorland differs from a natural ecosystem.

Answers on p. 130

Tropical biome

Equatorial rainforest

The equatorial rainforest **biome** is found within 10° of the equator in lowland regions in South and Central America, Central Africa and Indo-Malaysia. The most extensive area of rainforest is in the Amazon basin, where it covers 5.5 million km².

> **Biome** describes a large-scale ecosystem, often at a continental scale.

Climate

Equatorial rainforest is primarily a response to a hot and humid climate with:

- average annual temperatures between 25°C and 30°C
- little seasonal variation in temperature
- high average annual rainfall (>2,000 mm), with no dry season

High temperatures are a response to intense insolation. However, temperatures are slightly depressed by significant cloud cover, and are lower than in hot deserts in higher latitudes. Seasonal differences in temperature are small (2–3°C) because of (a) small seasonal variations in the sun's angle at noon (67°–90° at the equator), (b) day length being constant at around 12 hours.

Convectional rain falls all year round, though there are peaks around the equinox associated with movements of the ITCZ.

> **Typical mistake**
>
> Rainfall occurs at all seasons in the equatorial rainforest, but is not distributed uniformly. This is due to seasonal movements of the overhead sun which affect the position of the ITCZ. Shifting cultivators burn trash from forest clearings during drier months.

Soils

Heavy rainfall, which exceeds evapotranspiration, leaches minerals and nutrients from rainforest soils (**oxisols**). As a result soils are acidic and infertile. High temperatures and humid conditions cause rapid decay and the release of mineral nutrients from leaf litter and other dead organic matter. These minerals are quickly absorbed and recycled by shallow-rooted trees.

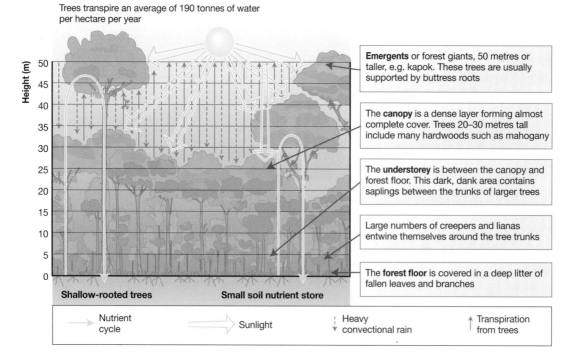

Figure 3.7 Rainforest structure and flows of energy, water and nutrients

Exam practice answers and quick quizzes at **www.therevisionbutton.co.uk/myrevisionnotes**

Plants and animals

Large evergreen trees dominate the rainforest. They create a variety of habitats and sustain the water cycle and the humid climate (Figure 3.7). Trees are festooned with lianas, vines, and epiphytes. Because the growing season lasts all year, plant productivity is high. Abundant moisture, warm conditions, climate stability (the rainforest is millions of years old) create a myriad of **ecological niches** which make the rainforest the most biodiverse ecosystem on the planet. Of the world's plant and animal species, 40% are in the tropical rainforest, and 1 km^2 of rainforest in Amazonia may contain up to 50,000 species of insect.

> **Typical mistake**
>
> Because rainforest vegetation is lush and primary productivity is high, it is often mistakenly assumed that rainforest soils are very fertile. In fact they are impoverished and contain only small stores of nutrients. The rainforest survives thanks to the rapid cycling of nutrients. Most nutrients in the rainforest biome are stored in the trunks and branches of trees.

Ecological responses to climate and soils

Revised

The main adaptations of plants and animals to the physical environment of the equatorial rainforest are described in Table 3.2.

Table 3.2 Adaptations of plants and animals to the rainforest environment

	Plant/animal adaptations
Light intensity	Only 1% of insolation reaches the forest floor. As a result the forest structure is layered. There are five layers: the supercanopy of emergent trees (40 m); the canopy (25 m); subdominant trees (10 m); small trees and tree ferns; and a sparse understorey of shrubs and herbs. Photosynthesis mainly occurs in the canopy. Flowers, fruits and leaves in the canopy attract most birds, mammals, amphibians/reptiles and insects. Many tall shallow-rooted trees have buttress stems for support. Typical canopy trees are branched, forming an inverted umbrella shape to capture the maximum sunlight. Many plants (e.g. epiphytes) grow on tree branches and stems.
Rainfall	Trees have pointed leaves with drip-tips, to shed potentially damaging rainfall.
Sunlight	Tree leaves are often leathery to protect against intense sunlight.
Seasonality	In the absence of seasons, trees are evergreen, flowering, fruiting and shedding leaves all year round.
Soil infertility	Lack of soil nutrients is compensated for by a rapid nutrient cycle. To facilitate rapid nutrient uptake, trees have dense networks of shallow roots. Decomposition by fungi, microorganisms and insects is rapid in the warm, humid environment of the forest floor.

Human activity and its impact

Revised

Human impact on the equatorial rainforest biome is proportional to the technology of a society. Indigenous people with low technology and small, stable populations have lived sustainably in the rainforest for millennia. During this time their impact on the ecosystem has been negligible. In contrast, advanced technological societies, particularly in the past 50 years, have had a devastating impact. According to the UN's Food and Agriculture Organisation (FAO) nearly 1 million km^2 of rainforest was destroyed every year between 2000 and 2005. Of this, 60% was primary forest. 200 years ago rainforest occupied 14% of the Earth's land surface. By 2010, **deforestation** had reduced this figure to just 6%.

Indigenous societies

Today only around 100 indigenous tribal groups, such as the Yanomani in Brazil and the Penan of Sarawak, remain in the rainforests. Most indigenous societies in rainforests have been either destroyed or changed irrevocably through contact with more technologically advanced cultures in the last century.

Indigenous rainforest cultures relied on hunting wild animals, gathering edible forest fruits and plants, and **shifting cultivation**. In ecological terms, this was a highly efficient and successful way of life. Shifting cultivation remains the only truly sustainable method of food crop production in the rainforest. Its sustainability is achieved by:

- temporary, rather than permanent cultivation of land. Given the infertility of rainforest soils, nutrient stores are depleted after just one or two harvests. As yields decline, farmers abandon their plots and make new clearings
- burning trash from forest clearings (branches, leaves, stems) which adds essential nutrients such as potassium and calcium to soils and destroys weeds
- **polyculture**, with a mixture of different crops sown in the ash in the clearings. This practice mimics the biodiversity of the forest, reduces the potential damage from plant pests and diseases, and by providing complete cover to the soils, reduces soil erosion by runoff

Technologically advanced societies

In contrast to indigenous cultures, most exploitation of rainforests by technologically advanced societies is unsustainable. At worst the impact has been outright destruction. Where development has stopped short of destruction, rainforest ecosystems have often been severely degraded. Unsustainable development is linked to economic activities such as farming, logging, dam building for HEP, mining and road construction (Table 3.3).

Table 3.3 Unsustainable resource development in the rainforest

Farming	Settlers clear small forest plots (often by burning) for permanent cultivation of food and cash crops (e.g. coffee). Declining yields due to soil exhaustion result in the abandonment of plots. They either become secondary forest, with impoverished biodiversity, or poor quality grassland for ranching enterprises. In parts of Amazonia, commercial soya production occupies large areas of former rainforest.
Logging	Logging companies extract valuable hardwood trees. In the process they destroy many other, non-commercial species and degrade the forest ecosystem. Further destruction occurs as roads are cut through the forest to provide access. Logging is often illegal and, without pro rata plantings of young trees, is unsustainable.
Dam building	Huge areas of forest are flooded by dams built to generate HEP. Construction also adds to deforestation.
Mining	Open-cast mining for minerals (e.g. iron ore, bauxite) results in localised destruction of the rainforest. Toxic elements and chemicals may pollute rivers and enter food chains.
Road building	Road building creates access to the rainforest for economic activities such as farming and mining, which greatly accelerates rates of development and deforestation.

Development issues Revised

Development of the equatorial rainforest biome raises a number of social, environmental and economic issues.

Social issues

Native people in Amazonia, the Congo basin and Indo-Malaysia have inhabited the rainforest for thousands of years. Today only 100 or so tribes survive, isolated from the modern world. Contact with technologically more advanced cultures has invariably proved disastrous. Indigenous people often have no resistance to common pathogens such as measles, chicken pox and flu. Meanwhile, deforestation by mining, logging and farming gradually destroys the forest resources that support them. There are moral issues here — should indigenous people be contacted or should they be left alone, free to pursue their traditional way of life and preserve their unique culture? Should vast areas of forest be preserved as native reservations for minority groups, or should they be developed to kick-start economic growth and benefit the majority of citizens?

Environmental issues

The equatorial rainforest sustains almost two-fifths of all plant and animal species. For both practical and aesthetic reasons this biodiversity should be protected. Many plant species have potential value either as medicines or as food crops. Others may become extinct before science discovers them. Locally, deforestation damages ecosystems through accelerated runoff, soil erosion and excessive sediment loads in rivers and streams. At a regional scale rainforests play a crucial role in the hydrological cycle. Without transpiration from forest trees, rainfall would decline, forests would disappear and entire ecosystems collapse. The world's rainforests are also important stores of carbon. According to NASA the destruction of tropical rainforest trees is responsible for 15–20% of annual global carbon emissions adding to problems of global warming and climate change.

Policies for sustainability

Rainforests can be protected by policies such as **debt exchange**. In return for writing-off a proportion of a country's international debt, that country agrees to conserve some of its rainforests. Agreements of this type have been popular between MEDCs and some Central and South American states. Also, in exchange for setting aside forest for carbon storage, countries with rainforests receive payments from MEDCs known as **carbon-offset** programmes. For example, in 2006 Bolivia received US$25 million for the sale of carbon credits earned by protecting an area of Amazon rainforest from logging. **Deforestation charges** levied against companies whose activities involve deforestation and **ecotourism charges** are other ways of protecting the rainforests.

Some rainforests in countries such as Costa Rica, Brazil and Indonesia have been designated as national parks. In Sumatra three national parks cover 25,000 km². As well as environmental protection, national parks also generate income from tourism to fund future conservation projects.

> ### Examiner's tip
> Exam questions dealing with development issues require an understanding of the economic, social, political and environmental arguments as well as the management responses designed to resolve these issues.

> ### Now test yourself
>
> 8 Describe the global distribution of the equatorial rainforest.
> 9 Explain why temperatures in the equatorial rainforest are often lower than temperatures in hot deserts in higher latitudes.
> 10 Why is there so little seasonal variation in temperature in the equatorial rainforest?
> 11 Examine the importance of rapid nutrient recycling to vegetation in the rainforest.
> 12 How do indigenous people exploit the rainforest sustainably?
> 13 Outline four causes of present-day deforestation in the equatorial rainforest.
>
> Answers on p. 130

Ecosystem issues on a local scale

Urbanisation and ecological change
Revised

Urbanisation often destroys natural ecosystems, replacing vegetation and soil with buildings, roads, tarmac and concrete. Less drastically, urbanisation alters ecosystems by modifying local energy exchanges, hydrology and soils. Compared with natural ecosystems, urban ecosystems have:

- lower biodiversity, with fewer **endemic** plant and animal species
- many alien or exotic species introduced either deliberately (e.g. cypress trees in gardens in the UK) or inadvertently (e.g. Japanese knotweed in the UK)
- reduced energy flows and therefore lower net primary productivity
- simpler structures, with shorter food webs and fewer ecological niches
- reduced nutrient cycling

> ### Typical mistake
> It is simplistic to argue that urbanisation merely destroys ecosystems. Urban ecosystems may be less biodiverse and less productive than most natural ecosystems, but towns and cities create new ecological niches and opportunities for wildlife.

However, urban ecosystems also create new ecological niches (e.g. nesting sites for peregrine falcons on tall buildings) and supplies of food and energy (e.g. from waste tips, dustbins, gardens). As a result, some animals such as house mice, red foxes, black-headed gulls and rats, which have successfully adapted to change, have now become 'pests'.

Urban niches

Urban environments provide a rich variety of habitats that benefit wildlife (Table 3.4).

Table 3.4 Wildlife and urban ecological niches

Gardens and parks	Gardens and parks provide cover for birds and mammals as woodland, shrub, grassland habitats. They also provide wetland such as ponds and lakes. Bird species that use feeders and nest boxes in garden/park habitats have increased their populations (e.g. blue tits, long-tailed tits, goldfinches). Swifts, swallows and martins nest in the eaves of buildings. Amphibians (e.g. frogs, newts) thrive in garden ponds. Garden plants (e.g. buddleia) provide nectar for butterflies, bees and other insects. Mammals successfully adapting to garden/park habitats include hedgehogs, rabbits, grey squirrels and foxes.
Churchyards	Overgrown and lightly managed churchyards provide habitats for lichens, wild flowers, insects, birds and small mammals (e.g. bats).
Wasteland	Wasteland provides opportunities for plant colonisation and succession. Early succession creates valuable cover for birds and mammals, as well as a rich food source. Old quarries provide nest sites for birds of prey. Mining and industrial spoil often favour plant species that cannot compete with more aggressive grasses (e.g. orchids).
Transport routes (roads, railways, canals)	Road and motorway verges, managed to restrict invasion by woody species (but otherwise free from human activity), support large populations of voles, shrews and other rodents, which in turn attract predators such as kestrels and barn owls. Old railway cuttings and embankments provide rich habitats, often protected as nature reserves. Canals provide habitats for fish and amphibians, and wetlands for aquatic plants, birds (e.g. water fowl) and rare mammals (e.g. water voles).

Ecosystems and the impact of new species

The introduction of new or alien species into ecosystems can have unforeseen and wide-reaching consequences. Given the complexity of food webs, any introduction of new species poses considerable risks.

Planned introduction — cane toads

In 1935 cane toads were introduced to Queensland, Australia, from Hawaii. It was hoped that they would predate and control beetles that damaged the sugar cane crop. Without natural predators and able to outbreed native frogs and toads, cane toads proved highly invasive and today occupy large parts of eastern Australia. In addition to beetles they consume native insects, arthropods, small mammals and birds, threatening biodiversity. Cane toads also emit venom through their skin which both deters and kills potential predators. Snakes, monitor lizards, even freshwater crocodiles that normally prey on frogs and toads are killed if they ingest cane toad venom. Cane toads have even colonised cities where they are a threat to pets and children. So far, eradication programmes have proved ineffective. While actions by community groups can reduce toad populations locally, strategies at the national scale now focus on containing the spread of the toads rather than eradication.

Unplanned introduction — mink

The American mink has become established in the British countryside during the past 60 years, having escaped from fur farms. The mink is highly adaptable — it is a strong swimmer and climber and preys upon small mammals, fish and ground-nesting birds such as ducks and moorhens. A highly successful predator and with few natural enemies in the UK, the mink outcompetes smaller carnivores such as stoats and weasels, and may be responsible for pushing water voles to the verge of extinction. Trapping programmes have had little impact nationally, but have been more effective locally in protecting endangered species such as water voles. However, there is some evidence that the return of otters to British rivers is helping to reduce mink numbers.

Examiner's tip

Try to convey a sense of the complexity of ecosystems in examination answers. Food webs are especially poorly understood. Thus human intervention, such as the introduction of new species or the eradication of predators, can have potentially devastating unforeseen effects.

Changes in the rural–urban fringe
Revised

Land-use changes in the **rural–urban fringe** have both negative and positive effects on ecosystems. **Urban sprawl** destroys natural habitats (woodland, meadows, ponds, etc.) with concomitant losses of wildlife. Planning strategies such as **green belts** have failed to stop urban sprawl. According to the Council for the Protection of Rural England roughly 110 km² of greenbelt land has been lost each year since 1997. Other threats to wildlife in the rural–urban fringe include dog walking, wind turbines, landfill sites, and vandalism.

More positively, many local authorities and other organisations have created conservation areas in the rural–urban fringe. Old mine workings have been reclaimed and revegetated in former coalfields; gravel pits and sewage beds have been converted to wetland habitats; country parks have been established; and nature reserves set up, managed by wildlife trusts, local authorities and other conservation bodies.

Typical mistake

Land-use change in the rural–urban fringe is not always damaging to wildlife. There are many examples of positive changes, creating new habitats and enriching biodiversity.

Case study Latterbarrow nature reserve, Cumbria

Wildlife Trust Cumbria Wildlife Trust (CWT) is a charity, set up in 1962. Its main brief is to protect and manage the wildlife in nature reserves in the county, largely through the work of volunteers. In total there are 46 Wildlife Trusts in the UK. The CWT currently has responsibility for 40 nature reserves throughout Cumbria.

Latterbarrow — habitat The Latterbarrow reserve, established in 1986, covers just 4.4 ha. Located in South Lakeland, between the Lake District hills and Morecambe Bay, it comprises an area of limestone grassland and woodland developed on thin, alkaline soils. Overgrown limestone pavements and limestone scars occupy part of the reserve. This type of limestone habitat, maintained by sheep grazing, was once widespread in South Lakeland. Latterbarrow is one of the few areas that survive today.

Biodiversity Given its size, Latterbarrow supports incredible biodiversity. Over 200 plant species are recorded in the reserve. There is a succession of flowers throughout the spring and summer, including seven

species of orchid, some of them at northern edge of their range in the British Isles. Other notable species include cowslips, primroses, rock rose and columbine. The reserve also supports a rich variety of insects, including moths and butterflies; birds, such as warblers, flycatchers and finches; and mammals, e.g. bats, foxes, stoats, rabbits and badgers.

Conservation and management Members of the CWT manage the reserve. The main task is to control the invasion of scrub such as blackthorn and brambles which threaten to shade out the grassland species. Much of this work is done by hand, though in the past livestock grazing would have maintained the open grassland. Today sheep are used on the reserve for part of the year, and occasionally, when scrub invasion becomes more severe, cattle are grazed over winter. The site is easily accessible to visitors, and trampling and dog walking are problems. Notice boards provide educational information to visitors.

Ecosystem issues on a global scale

Human activity, biodiversity and sustainability

Revised

In the past 50 years the consumption of natural resources, both in absolute and per capita terms, has grown massively. Ever-increasing demands for food, energy, minerals, water and timber have led to the unsustainable exploitation of ecosystems. Evidence of this, such as deforestation, **desertification**, **land degradation**, overfishing and climate change is found on a global scale. The principal drivers are:

- massive world population growth, from 2.5 billion in 1950 to 7 billion in 2012, increasing absolute demand for resources

- rising standards of living, increasing per capita consumption of natural resources — a process that has accelerated in the past 20 years with the growth of **emerging economies** such as China and India

Typical mistake

Desertification and land degradation are different processes. Desertification describes processes that reduce biological activity to a point where desert-like conditions prevail. Land degradation is more widespread and less severe. It describes a deterioration in the quality of farmland due to processes such as salinisation, overcultivation, soil erosion, etc.

Managing fragile environments

Revised

Environments such as **drylands** (arid and semi-arid areas) and high mountains are easily damaged and degraded by human activities. These environments are fragile and damage may be irreversible because of:

- high degrees of endemism and lack of biodiversity

- slow rates of soil formation and plant growth

- sparse vegetation cover that is easily destroyed, with steep slopes (mountains) and aridity (deserts) leaving soils vulnerable to erosion

Case study Combating desertification and land degradation: The Korqin sandy lands and China's Great Green Wall

Background Environmental degradation leading to desertification affects one-third of China's total land area. China loses 5 billion tonnes of topsoil to erosion every year. Most seriously affected are the arid and semi-arid regions of northern China.

The Korqin sandy lands of northern China are especially fragile and vulnerable to overexploitation. Overgrazing and clearing trees for timber and farmland (both driven by human population growth), has exposed the sandy soils to erosion. Strong winds, responsible for violent dust storms, are the main agents of erosion. The result is large expanses of degraded and desertified land. Excessive extraction of groundwater, which has lowered water tables, and poor drainage leading to soil salinity, have also contributed to environmental degradation.

Management responses In 1978 China adopted an ambitious programme to combat land degradation and desertification in its northern provinces. The scheme, popularly known as the Great Green Wall, is a massive reafforestation programme. The aim is to establish 350,000 km^2 of plantation forests and shelterbelts by 2050. So far, 130,000 km^2 have been planted.

In Korqin the programme aims to protect farmland against wind erosion, restore soil fertility and improve the well-being of local people. Plantations and shelterbelts have been planted with native poplar trees, resistant to drought and frost. They also provide a sustainable source of timber.

But the scheme is not just about reafforestation. There is an emphasis on conservation, with the introduction of sustainable cultivation involving:

- the recycling of organic material to the soil

- integrating tree crops with pasture and cash crops (agro-forestry)

- planting tree species that provide fodder and timber and improve soil fertility

Meanwhile, the controlled management of grazing lands (i.e. keeping within the land's carrying capacity) is being introduced for the first time.

Case study Conservation and sustainable management of ecological resources: Annapurna, Nepal

Background The Annapurna region of Nepal is a spectacular mountain environment in the Himalaya reaching over 8,000 m. As well as classic alpine scenery there are deep gorges, glaciers and icefields. The snowline varies between 4,000 and 5,000 m. This high-energy environment is associated with rapid river incision, landslides and mudslides — processes that are particularly active during the monsoon rains. Climate varies with altitude, creating a mosaic of habitats with many endemic plant and animal species.

Ecological problems Population growth and poverty have encouraged overgrazing, the intensification of farming and rising demand for firewood, fodder and timber, causing deforestation and soil erosion. Ecotourism has added to environmental pressures. The region has 60,000 visitors a year — mainly trekkers. They erode mountain trails, burn firewood from the forests and their pack animals overgraze areas close to trails. Organic waste generated by tourists pollutes mountain rivers and streams. Solid waste is a problem along popular trails.

Management responses The Annapurna Conservation Area (ACA) was given protected status in 1986 and a management plan — the Annapurna Conservation Area Project (ACAP) — was set up to provide a sustainable plan to meet the needs of local people, tourists and nature conservation. The ACAP gets its income from tourists, who are charged a visitor fee, the World Wildlife Fund for Nature and foreign aid. The ACAP supports projects such as reafforestation and the conservation of water, soil and wildlife resources. These programmes not only help conservation but also provide employment for local people who work as trekking guides, forest guards, tree nursery personnel, etc. Conservation is thus seen as a way of raising the incomes of local people. As such it receives their support because they see an economic advantage in protecting the environment and wildlife. Meanwhile, external agencies have advised on better livestock management, organised tree planting and promoted secondary forest regeneration in areas previously cleared for farming.

Surveys in the ACA have shown significant increases in the forest area, biodiversity and wild animal populations. Further progress will be made as family planning programmes begin to take effect and kerosene stoves and small HEP generators replace firewood as the main fuel used by local people.

Now test yourself

14 Describe how urban areas can create new opportunities for wildlife.

15 Why is the introduction of new (alien) species into ecosystems often disastrous?

16 Give two reasons why biodiversity and the sustainability of ecosystems are under threat.

17 What makes some environments particularly fragile and easily degraded by human activities?

18 What can be done to protect fragile environments from overexploitation and degradation?

Answers on p. 131

Check your understanding

1 Explain why energy declines with distance from the site of primary production in food webs and food chains.

2 Describe and explain the process of primary succession on tidal mudflats.

3 Describe and explain the vertical structure of the temperate deciduous forest.

4 Suggest reasons for the extraordinary biodiversity of the equatorial rainforest.

5 In what ways does shifting cultivation in the rainforest mimic the natural ecosystem?

6 To what extent is contemporary economic development in the equatorial rainforest unsustainable?

Answers on p. 131

Exam practice

Section A

1 (a) Study Figure 3.3 which shows a food chain in an oak woodland. Describe and comment on the main characteristics of the food chain in Figure 3.3. [7]

    (b) Describe and explain the changes associated with ecological succession in one ecosystem you have studied. [8]

    (c) With reference to a specific example, discuss the effect of human activity on ecological succession and climax vegetation. [10]

2 (a) Study Figure 3.7 which shows the structure of the tropical rainforest. Comment on the structure of the tropical rainforest in Figure 3.7. [7]

    (b) Describe and explain the impact of indigenous cultures on the tropical rainforest biome. [8]

    (c) Using examples, assess the sustainability of contemporary economic development in the world's tropical rainforests. [10]

Section C

3 Assess the impact of human activity and development on one tropical biome that you have studied. [40]

4 With reference to development in one or more fragile environments, discuss the extent to which sustainability has been achieved. [40]

Answers and quick quiz 3 online

Online

Examiner's summary

✔ Understand the functioning of ecosystems requires an appreciation of their holistic character (i.e. interrelatedness). The holistic nature of ecosystems is controlled by energy flows through food webs, and nutrient recycling.

✔ Ecological succession occurs gradually in stages and eventually may result in a stable, climatic climax form. However, climatic climax may be permanently arrested by the influence of other factors, e.g. human activity, soils, relief.

✔ Detailed knowledge and understanding often differentiate good from modest answers, e.g. that equatorial rainforest often has two rainfall maxima and a short dry season; that rainforest soils store few plant nutrients, resulting in the failure of permanent cultivation.

✔ The success of a farming system can be measured by its sustainability, and its impact on natural ecosystems.

✔ Before adopting a value position, exam answers that discuss issues and conflicts must show a balanced appreciation of the arguments of the different protagonists.

✔ As well as learning the background to issues and conflicts (e.g. causes, impacts), responses (i.e. attempts at planning and management) should be known.

✔ In questions requiring extended answers, detailed and place-specific case studies must be learned. Generalised answers, even if detailed and accurate, are unlikely to secure more than a modest grade at A2.

4 World cities

Global patterns

Distribution and types of large cities

Revised

Large cities dominate the global **urban hierarchy**. There are three groups of large cities:

- **millionaire cities**, with populations of 1 million and above
- **mega cities**, with populations of 10 million and above
- **world cities**, which have large populations *and* provide important services to the international economy

There are no standard criteria for defining urban areas and urban populations. Thus, estimates of the numbers of different types of large city vary widely. In 2009, according to UN estimates, there were 443 millionaire cities, of which 21 were mega cities. Around one-fifth of the world's millionaire cities are in China, and more than 70% are in the developing world.

World cities

What distinguishes world cities from mega cities is their global influence. While mega cities perform services at regional and national scales, world cities are the command and control centres of the international economy. In 2010 the three top-ranking world cities were New York, London and Tokyo (Table 4.1). Typically, world cities are: leading business centres and headquarter locations of major TNCs (transnational corporations); global service centres for finance, banking, accounting, management consultancy, law and advertising (i.e. **producer services**); and important information centres and transport hubs. World cities also have highly educated workforces, world-class universities, and host a wide range of international cultural, sporting and political events.

Typical mistake

It is often forgotten that three in every four urban dwellers live in the developing world and that this proportion will increase steadily for at least the next 30 or 40 years.

Table 4.1 Top ten world cities 2010

Rank	City
1	New York
2	London
3	Tokyo
4	Paris
5	Hong Kong
6	Chicago
7	Los Angeles
8	Singapore
9	Sydney
10	Seoul

Source: *Foreign Policy* magazine with AT Kearney and the Chicago Council of Global Affairs, 2010

Economic development and urban change

Revised

Globalisation and rapid economic development in the past 30 years in Asia, especially in emerging economies such as China, have triggered massive urban growth. For much of this time, growth in China's economy has averaged more than 10% a year. Today, the Chinese economy is the second largest in the world after the USA and accounts for 9% of total international trade. China's economic growth has been led by exports of manufactured goods, though as living standards rise China's domestic market will become increasingly important.

Economic development in China has been accompanied by spectacular urban growth. In 1980 there were 19 Chinese cities with populations in excess of 1 million. In 2010 there were 88 (UN estimate). This growth, driven by rural–urban migration, has been responsible for a major shift in population distribution. In 1980 just over one-quarter of China's population lived in towns and cities. By 2015 the majority of Chinese will be urban dwellers.

Contemporary urbanisation processes

Urbanisation

Urbanisation is the proportion of urban dwellers in a country or region. **Urban growth** describes an absolute increase in the number of urban dwellers.

When urbanisation increases there is a relative shift of population from rural to urban areas. Such a shift often occurs when:

- **rural–urban migration** produces a **net migrational gain** in urban areas and a **net migrational loss** in rural areas
- rates of natural population change (i.e. annual difference between births and deaths) are greater in towns and cities than in the countryside

Typical mistake

Because urbanisation refers to the proportion of urban dwellers in a country or region, urbanisation can increase even when the absolute number of people in urban areas falls.

Rapid urbanisation has occurred at the global scale in the past 50 years. By 2010 half the world's population lived in towns and cities. Europe, North America, South America and Oceania are the most urbanised continents; Africa and Asia are least urbanised. Although only 43% of Asia's population lives in towns and cities, the absolute number of urban dwellers in Asia (around 1.7 billion) is the largest of any continent.

Urbanisation in the developing world has been driven by two sets of factors. First, the concentration of investment and economic activity (which provide wealth and employment) in towns and cities. Second, the poverty and lack of investment in the countryside.

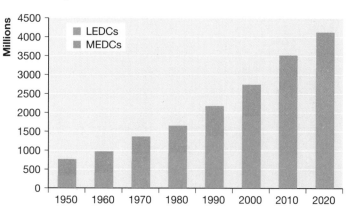

Figure 4.1 Urban population growth in MEDCs and LEDCs, 1950–2020

Suburbanisation

Suburbanisation describes the growth of suburbs through the decentralisation of population, industry and other business activities. In the UK, large-scale suburbanisation was essentially a twentieth-century phenomenon, though it started in the late nineteenth century. Suburban development mainly comprised low-density housing and industry on **greenfield sites**. The causes of suburbanisation in the UK were:

- expanding urban populations and a shortage of housing
- rising standards of living with demands for higher quality housing at lower density

- the availability of building land around the edges of towns and cities
- the growth of local authority planning in the 1920s and 1930s and strategies to ensure orderly urban growth
- improvements in public transport (e.g. electric trams, suburban railways, buses) making it easier and cheaper to access the suburbs

The consequences of suburbanisation were:

- **urban sprawl** and the loss of countryside and amenity (e.g. farmland, wildlife) in the rural–urban fringe. In England between 1997 and 2010, around 150 km^2 of **greenbelt** were lost to urban development and nearly 45,000 houses built
- longer **journeys to work**, which in the past 50 years have increasingly caused problems of traffic congestion and pollution at peak times
- the demographic and economic decline of central urban areas — as population has moved to the suburbs, many services (e.g. retailing) and businesses (e.g. offices) have followed. Commercial activities have also been attracted to the suburbs which offer greater accessibility, more space, and lower rents, compared with more central urban locations (e.g. CBD (central business districts), inner city)

> **Examiner's tip**
>
> A useful way of learning the causes of suburbanisation is to divide them into 'push' and 'pull' factors. Push factors are problems within central areas (e.g. lack of space, poor quality housing) that encourage people and businesses to leave. Pull factors are the attractions of the suburbs (e.g. space for expansion, better quality housing, etc.)

Counterurbanisation Revised

Counterurbanisation is an increase in the proportion of a country's or region's population living in rural and semi-rural areas. In most instances counterurbanisation is due to **urban–rural migration**. Counterurbanisation was an important process in the changing geography of population in the UK between 1970 and 1990. In the UK it mainly involved urban dwellers migrating from metropolitan counties such as Greater London to adjacent rural counties and districts within commuting distance, such as Berkshire. Counterurbanisation is also influenced by retirement of higher-income groups to scenic coastal areas (e.g. Dorset) and national parks (e.g. Lake District). The causes of counterurbanisation are:

- the perceived better quality of life in rural areas and small towns (e.g. less polluted environment, less crime, better schools, etc.)
- motorways, modern roads and rail connections which make commuting easier
- dissatisfaction with the quality of life in large cities
- lower house prices in rural areas (e.g. Suffolk) than in large metropolitan areas (e.g. London)
- telecoms advances that make it possible to work from home (e.g. internet, e-mail)

> **Typical mistake**
>
> A common misconception is that counterurbanisation is due to a relative shift of population from cities to rural settlements such as hamlets and villages. In the UK, counterurbanisation mainly involves urban dwellers moving out, beyond the green belt, to smaller urban settlements in rural districts.

Reurbanisation Revised

Reurbanisation describes the renaissance of urban life in the past 30 years or so, in some of the most run-down and neglected parts of cities in Europe and North America. It reflects a renewed desire, particularly among young, professional groups, to live in or close to the city centres. The process involves centralisation and is therefore the opposite of suburbanisation. Reurbanisation is often accompanied by extensive renovation of properties, which moves house prices and rents upmarket. It also creates a demand for new neighbourhood businesses such as cafes, restaurants, delicatessens, wine bars, etc. This upgrading process by higher-income groups is known as **gentrification**. Examples include many parts of London (e.g. Docklands, Islington and Clapham), Chicago's Near West Side, and Puerto Madero in Buenos Aires.

Reurbanisation is driven by:

- a preference to live closer to workplaces in city centres to reduce the time and cost of commuting
- the attraction of new leisure services and night life in city centres (e.g. restaurants, theatres, cinemas, clubs) particularly for younger age groups with higher incomes
- government planning projects in inner city locations (e.g. docklands) designed to reclaim and revitalise **brownfield sites**, regenerate inner cities, and reduce congestion and pollution caused by commuting

> **Typical mistake**
>
> Reurbanisation is less about a migration of people from the suburbs to inner city locations and more about the residential preferences and life styles of younger, professional groups.

Planning and management issues

Revised ☐

Contemporary urbanisation processes often give rise to a range of planning and management issues. These are summarised in Table 4.2.

Table 4.2 Contemporary urbanisation and planning and management issues

Process	Issue
Urbanisation	Urbanisation is mainly a problem in cities in LEDCs. The speed and scale of urbanisation creates issues of poor quality housing and housing shortages, inadequate services, environmental pollution, unemployment, etc. The growth of informal settlements raises issues of sustainability, policing, international image, etc.
Suburbanisation	Suburbanisation in MEDCs puts pressure on the rural–urban fringe. Housing projects on greenbelt land are often resisted by suburban residents who see them as a threat to their quality of life.
Counterurbanisation	Counterurbanisation by commuters creates longer journeys to work, adding to road congestion and pollution. It can undermine rural economies, by reducing the demand for public transport, education and retail services, and cause hardship for old people and those without independent transport. Large influxes of commuters and retirees can also damage the social fabric of rural communities and can price local families out of the housing market.
Reurbanisation	Reurbanisation can create inequalities. Lower income families may be excluded from living in gentrified suburbs because of increased prices and rents. New housing developments in renovated districts may include insufficient affordable housing.

Case study — Urbanisation in Nairobi, Kenya: causes and impact

Background Nairobi is the largest city in Kenya and one of the fastest growing in Africa. Its population increased from 137,000 in 1950, to 3.5 million in 2010. By 2025 its population is expected to reach 6.25 million. Like most countries in sub-Saharan Africa, urbanisation has been rapid in the past 60 years. In 1950 only 5.6% of Kenyans were urban dwellers, today the proportion is 23%.

Causes of urbanisation The primary demographic cause of urbanisation is rural–urban migration. Although fertility and natural increase rates are higher among the rural population, Kenya's rural population only expanded three-fold between 1950 and 2010 compared with the urban population which grew by a factor of 10. The primary economic cause of rural–urban migration is poverty in the countryside. Although Kenya is a poor country (GNI per capita $US770 in 2009) the incidence of poverty is far higher in rural areas than in the towns and cities. The government has also concentrated most investment and economic development in the past 20–30 years in the major cities of Nairobi and Mombasa.

Impact of urbanisation Massive and rapid urban growth has created huge economic and social problems. In Nairobi more than half the population lives in slums such as Kibera, which are overcrowded and insanitary. A typical slum dwelling is a one room shack (3–6 m²), housing five or six people. Overcrowding promotes the spread of water-borne diseases, diarrhoea and TB.

In Kibera open sewers flow along streets that turn to mud after rain, and there are just 600 pit latrines for 1 million people. Most households have no piped water and rely on standpipes, tanks or water vendors. There is no organised rubbish collection; garbage is thrown everywhere and only 20% of households are connected legally to the electricity grid. Crime rates are high, made worse by the absence of street lighting.

Jobs in the formal sector (e.g. in construction) are scarce and are located in Nairobi's city centre and the commercial/industrial spine of Mombasa Road. Many people cannot afford transport and walk long distances to work each day. Because unemployment is not an option (there is no welfare system) people rely on the informal sector and self-help (e.g. work such as street vending, repairing and recycling, prostitution, etc.).

Case study **Suburbanisation and greenfield development in West Yorkshire**

Background The population of Bradford is expected to grow by 140,000 between 2009 and 2026. This is due to the effects of immigration and the city's youthful age structure. One effect will be a large increase in the demand for affordable housing (50,000 extra homes needed 2009–2026). Some of this new housing will be on brownfield sites within the city. However, there will still be a shortfall, which means building on greenfield sites that are (or were) protected by greenbelt legislation. Development on greenfield sites is unpopular with local residents who invariably bear some of the costs.

Location Menston is a commuter exurb located in the Bradford Metropolitan District. It is a commuter settlement mainly for Leeds, and to a lesser extent for Bradford and Harrogate. Menston is one of the most prosperous areas in the Bradford Metro and ranks among the 3% least deprived super output areas in England. Planners have approved the building of 300 houses by Barratt Homes and Taylor Wimpey on two greenfield sites. The sites are currently used as permanent pasture for grazing sheep and cattle.

Issue Local residents formed a pressure group to campaign against the plan. Their opposition gained strong support from the local community. In a referendum in 2011, 98% of residents who voted (i.e. half of all residents) were against the plan. However, the results were not binding on the council. The main opposition arguments are: the proposal will expand the built area of Menston with loss of countryside, biodiversity and amenity; increased traffic on local roads and congestion on the busy A65; increased pressure on Leeds train services which are already overcrowded at peak times; pressure on local services, especially primary schools and healthcare.

However, planners argued that they were unable to meet all of the demand for new housing by developing brownfield sites in Bradford. They maintained Menston has excellent rail connections to Leeds and Bradford and that using the rail network is environmentally less polluting, will take traffic off the roads, and is sustainable. The sites in question were removed from green belt status in 2004 in anticipation of development, therefore pre-empting the green belt issue.

Now test yourself

1 What is (a) a mega city, (b) a world city?
2 How does globalisation encourage the growth of large towns and cities?
3 Explain how it is possible for urban growth to occur without urbanisation in a country.
4 Outline two demographic factors responsible for rapid urban growth in LEDCs.
5 What is the difference between a greenfield and a brownfield site?
6 What is the difference between urbanisation and counterurbanisation?

Answers on p. 131

Urban decline and regeneration

Urban decline ————————————————————————————————— Revised ☐

Many cities in old industrial regions in MEDCs have experienced steep economic decline in the past 40 or 50 years. Cities in the USA's Rust Belt, such as Detroit, Buffalo, Cleveland, Philadelphia and Baltimore, are among those hardest hit. A similar decline has affected many towns and cities in the UK, particularly in the North and Midlands. Urban decline is associated with:

- the collapse of traditional manufacturing industries such as steel making, heavy engineering and automobiles (**deindustrialisation**)
- the creation of large swathes of derelict land, formerly occupied by factories, docks and mines
- high rates of unemployment and a workforce lacking in skills that are relevant to modern service economies
- high incidences of poverty, **multiple deprivation** and **social exclusion**
- migration either within the metropolitan region or to other parts of the country, resulting in population decline (Table 4.3)
- the decline of retailing and other services as demand falls

Table 4.3 Population change in the USA's Rust Belt: 1990–2010

	1990	2010	%change
Detroit	1,028,000	713,777	−30.6
Cleveland	505,616	396,815	−21.5
Buffalo	328,123	261,310	−20.3

The inner areas of cities have been most affected by urban decline, as businesses and jobs have disappeared and middle-income families have moved out. In the worst cases, property prices have collapsed, leaving abandoned houses and streets.

Urban decline in MEDCs has been most severe in cities overdependent on traditional manufacturing industries. Lacking economic **diversification**, these cities have been extremely vulnerable to downturns in the global economy and to competition from lower-cost producers. New technologies have also made manufacturing less labour intensive, reducing the demand for employment and adding to decline. For example, 30 years ago the US automobile town of Flint in Michigan state employed 79,000 workers in car making. Today that number is down to 6,000 and unemployment stands at 20%.

Typical mistake

You should appreciate that urban decline has social and environmental, as well as economic dimensions.

Examiner's tip

In addressing the issue of urban decline in extended-answer questions, a useful distinction can be made between decline at intra-urban and inter-urban scales.

Urban regeneration

Revised

Urban regeneration attempts to reverse decline by improving the physical fabric and economy of decaying urban areas. Gentrification is a type of urban regeneration that relies on private capital. More often, urban regeneration is driven by governments. In the UK, earlier schemes in the 1980s were property-led and focused mainly on physical regeneration. More recent programmes have a wider brief that includes social and economic regeneration.

Property-led regeneration

In the UK in the 1980s and 1990s the government favoured **property-led regeneration**. Its flagship scheme was Urban Development Corporations (UDCs). Twelve UDCs were set up, covering some of the worst examples of urban decay, such as London's Docklands, the Don Valley in Sheffield and Cardiff Bay. The UDCs remit was to acquire derelict land, restore it for development and sell it on to private business. Essentially public money was used to prime-pump private investment. For example, in London's Docklands, an initial government spend of £1.86 billion had attracted £6.67 billion of investment from the private sector by 1999.

In total the UDCs delivered 27,000 new homes and provided office and factory space covering 5.4 million m². They also created around 150,000 jobs. Criticism of the UDCs focused on the relatively small proportion of new jobs that went to local residents, that generous subsidies offered by UDCs simply diverted investment away from other urban sites, and that the overall cost to the taxpayer (£3.5 billion) was too high.

Partnership schemes

More recent regeneration projects have aimed to achieve social and economic, as well as physical regeneration. These projects are usually **partnership schemes**, involving the public sector (government) and the private sector (business).

New East Manchester (NEM)

NEM is a partnership scheme between Manchester City Council, the North West Development Agency and the Homes and Communities Agency. Set up in 1999, NEM is one of the largest regeneration programmes in the UK. Its aim is to transform the social, economic and physical environment of a 20 km² tract of inner Manchester — an area beset with multiple problems such as poverty, unemployment, poor housing and industrial decline. The NEM's strategy is to attract new businesses, rebuild the area's economic base, help local residents get work and improve the physical environment.

Centrepiece of the scheme is the Etihad football stadium (site of the 2002 Commonwealth Games and home to Manchester City Football Club). Between 2000 and 2010 NEM built 5,000 new houses (including many affordable homes) and improved nearly 6,500 others; constructed three new shopping centres along with health centres and children's centres; and established parks, community gardens and sports facilities. By providing 200,000 m² of commercial floorspace, the NEM has strengthened the district's economic base. In total, 6,200 residents have been assisted into employment, 600 people retrained, and 180 have started their own businesses. Local retailers such as Tesco, Asda and Matalan have recruited and trained over 1,000 local people and the proportion of working residents claiming benefits has fallen from 40% to 33%. Three new high schools have helped to raise educational standards among school leavers.

> **Examiner's tip**
> You must understand the nature of urban regeneration schemes, *and* be able to evaluate their impact.

Retailing and other services

Decentralisation trends
Revised

The main geographical trend in urban retailing and other services in MEDCs in recent decades has been decentralisation.

Fifty years ago, high-order **comparison goods and services** were almost exclusively concentrated in **central business districts (CBDs)**. Demand for lower-order, **convenience goods and services** was provided by smaller suburban shopping centres. The result was an urban shopping **hierarchy**. At the top was the CBD, supplying high-order goods and services for the city and its surrounding region. Larger **district centres** in the suburbs, often located along major roads or at important road junctions, provided lower-order goods and services for several thousand households. At the base of the hierarchy were dozens of small **neighbourhood centres**, serving a few hundred households in surrounding streets.

> **Examiner's tip**
> Accurate and informed descriptions of urban service centres and hierarchies should demonstrate understanding of the key concepts of threshold and range.

The CBD

Decentralisation of retailing activities has changed the urban shopping hierarchy. The CBD has lost its dominance as the pre-eminent comparison retailer, as a wide range of stores — furniture, carpets, electrical goods, clothing and many others — have located in the suburbs. Meanwhile, **service retailing** has become more important in the CBD, compensating for the decline of **product retailing**. Increasingly CBDs are 'event destinations' catering for leisure and recreational activities, with cafés, restaurants, bars and clubs, as well as longer-established cinemas, theatres and galleries. Many CBDs have also lost large numbers of office jobs to business parks and office parks in the suburbs.

The suburbs

Retailing in the suburbs has also undergone a transformation. Since the 1970s planned shopping centres, comprising large supermarkets and comparison stores operated by major multiple retailers such as Argos, Curry's and PC World, have appeared at strategic transport hubs in the suburbs. The largest centres are equivalent in size to the central shopping areas of major cities. They include **regional shopping centres** such as Bluewater at Dartford and Meadowhall in Sheffield.

Meanwhile, thousands of small, independent retailers in suburban shopping centres, particularly those selling food, have disappeared. A net loss of 10,000 shops is forecast in the UK in 2011, the majority of which will be small independent businesses in the suburbs. Service retailers, on the other hand, such as hairdressers, beauty care and takeaway outlets, increasingly dominate suburban shopping centres, where they have replaced traditional product retailers such as butchers, greengrocers and newsagents.

> **Typical mistake**
>
> Decentralisation as a process does not usually mean that an urban retailer transfers a store in the CBD to the suburbs. Rather the shift in favour of the suburbs is the result of new investment being channelled there, rather than to the CBD.

Causes and impacts of decentralisation

Revised

The main reasons for the decentralisation of retailing and offices are:

● the migration of urban dwellers from inner urban areas to the suburbs, particularly in the past 60 years or so — businesses have followed their customers and workforces

● changes in the nature of retailing — the development of large supermarkets and other superstores benefiting from economies of scale and geared towards car-borne shoppers. These units need large sites for single-storey buildings and extensive parking space

● traffic congestion in the CBD and problems of access (deliveries by road, access for customers and workers)

● high cost of renting floorspace in the central parts of cities

● the policies of local authority planners, which have often sanctioned the growth of out-of-centre retailing and business parks

Decentralisation has had both positive and negative impacts. The positive impacts are:

● suburban populations have easier access to jobs, and to food, clothing, domestic electrical equipment and in many instances lower prices (some savings in scale economies are passed to consumers)

● deliveries to suburban retailers via peripheral ring roads, motorways and dual carriageways (often from large storage depots alongside motorways)

are more efficient, and help to reduce congestion in city centres. Distances travelled to work and shop are reduced, with environmental benefits

Among the negative effects of decentralisation are:

- widening inequality in access to retail services, especially food (i.e. quality and range) because decentralisation favours car-owning middle- and higher-income groups, it discriminates against low-income groups in inner city locations, and carless and older age groups who cannot easily access retail parks in the suburbs
- the closure of small independent food stores (unable to compete with supermarkets) means that fresh foods are less likely to be available in local, neighbourhood centres
- the decline of product retailing in city centres, leading to many vacant units and the growth of poundshops, pawnshops and charity stores, threatens the commercial health and vibrancy of the CBD
- the unsustainable growth of journeys by car for shopping trips in the suburbs increases carbon emissions, traffic densities and congestion
- inner city residents have fewer job opportunities in the CBD

> **Examiner's tip**
>
> Discussion of the effects of decentralisation could be structured in terms of social, economic and environmental impacts.

Case study Out-of-town retailing: Trafford Centre

Location and access Situated 7 km west of Manchester city centre between junctions 9 and 10 on the M60. Easily accessible from the M61, M62, M6 and M56. Metrolink will be extended to the Trafford Centre by 2016. There are 10,000 free parking spaces on site. 5.3 million people live within 45 minutes' drive-time.

Background Completed in 1998 on a brownfield site, the Trafford Centre is the second largest regional shopping centre in the UK. The main complex has 177,000 m² of floorspace, more than 200 shops, 60 cafés and restaurants and a multiscreen cinema. The anchor tenants are John Lewis, Debenhams, Selfridges and M&S. Around 30 million shoppers a year visit the Centre and the average spend per visit is £100. 8,000 people work at the Trafford Centre. The success of the Trafford Centre has attracted further

investment around the site, including the Barton Square expansion, the Legoland Discovery Centre and Trafford Quays leisure village.

Issues The Trafford Centre appears to have had little adverse impact on retailing in Manchester's CBD and in larger towns in Greater Manchester, such as Bolton and Wigan. However, it has hit smaller towns, such as Altrincham and Eccles, hard. In 2011 one-third of Altrincham's shop units were vacant — the highest proportion in the UK. The Trafford Centre provides an attractive alternative to Altrincham town centre owing to its proximity (12 km), greater retail choice (including leisure and catering attractions) and free parking. Congestion on the M60 at junctions 9 and 10, with long tailbacks at peak times, is a serious issue for the Trafford Centre.

Now test yourself

7 What is deindustrialisation?

8 Why has deindustrialisation had the most severe impact in cities lacking economic diversification?

9 What is property-led urban regeneration?

10 How has urban decentralisation increased inequality among city residents?

Answers on p. 132

The redevelopment of urban centres

Revised

Investment in city centre shopping malls, pedestrianisation and the growth of service retailing (e.g. clubs, bars, restaurants, cafés and hotels) have reinvigorated and broadened the appeal of British city centres. In addition to employment and retailing, city centres such as Newcastle, Manchester and Leeds have developed a distinctive culture, with nightlife that attracts tourists, students and other visitors as well as local people. After decades of decline, people have begun to move back into British city centres. Thousands of new apartments have been built and old office blocks, mills and warehouses converted for residential use. In 1990 only 1,000 people lived in Manchester's city centre. By 2005 this figure had risen to 20,000. Local authorities have encouraged this movement, which has brought in investment, regenerated run-down areas and has helped to create a vibrant city centre. It also reduces journeys to work, helping to make cities more sustainable. Recentralisation has been led by young couples and singles (half of all apartments in central Manchester are single occupancy). The attractions are lifestyle, leisure and cultural facilities, nightlife and proximity to work in the centre.

Typical mistake

The retail functions of the CBDs of British cities have changed in the past three or four decades. The CBD is no longer the exclusive centre of high-order product retailing, while service retailing, geared to recreation and leisure, has become relatively much more important. The latter trend partly reflects the growing popularity of city centres as places to live.

Case study The redevelopment of Birmingham city centre

Background Birmingham, with a population of more than 1 million, is at the centre of a region — the West Midlands — of 2.5 million people. The city centre, largely rebuilt in the 1950s and 1960s, was in urgent need of regeneration by the 1980s. Redevelopment in the 1960s created a bland and dreary urban environment dominated by brutal architecture and motor vehicle traffic. The unattractive environment was epitomised by the 1960s buildings such as New Street Station, the Bull Ring and the Rotunda. An inner ring road (also built in the 1960s) sliced through the centre, fragmenting the urban structure. Few people lived in the centre.

Existing regeneration Today the city centre is undergoing a renaissance as it is transformed into a place for tourism, education, arts, international conferences, state-of-the-art shopping and living. Regeneration began in the early 1990s with the completion of the Symphony Hall, the International Convention Centre and the National Indoor Arena. Retailing received a major boost in 2003 with the demolition of the old Bull Ring and its replacement with a £500 million award-winning shopping centre. This development, with its 160 shops and 25 restaurants has helped propel Birmingham to third place in the UK retail hierarchy (2011). Several new urban squares such as Victoria

Square and Centenary Square have added variety to the urban structure. Brindley Place, formerly a run-down industrial site, has been successfully regenerated as a canal-side development with apartments, offices, restaurants and a gallery. Meanwhile the city centre has been made more pedestrian-friendly, with interconnecting and vehicle-free streets and the demolition of the isolating, elevated section of the inner ring road.

Future regeneration Future regeneration is based on Birmingham's Big City Plan. Launched in 2008 it aims (over the next 20 years) to coordinate further improvements in the physical, cultural and economic redevelopment of the city centre. This £10 billion project should create 50,000 new jobs and 5,000 new homes. Its centrepiece is a £600 million redevelopment of New Street Station and the surrounding area (e.g. St Martin's Square). The remodelled station will double passenger capacity and reduce road congestion in the centre. Seven quarters in the city centre have been defined, including the historic Gun and Jewellery Quarters. Development will enhance their distinctiveness. The Eastside Quarter will become a cultural and educational area, as well as providing new housing, shops, leisure and business enterprises.

Sustainability issues in urban areas

Waste management Revised

The disposal of huge quantities of solid waste generated by towns and cities presents major environmental problems. London generates 4.4 million tonnes of municipal household waste a year. Most of this waste is disposed of in 18 **landfill** sites — some more than 120 km from the capital. Landfill is an environmentally unsatisfactory method of waste disposal for three reasons:

- leakage of toxic chemicals into the environment
- emissions of GHGs such as methane and carbon dioxide

- loss of countryside, amenity and brownfield sites such as old quarries, which could be used for other purposes in crowded southeast England

Recycling

The UK is rapidly running out of landfill space and needs to find an alternative in the next few years in order to meet EU directives. One approach is to extend **recycling** schemes and reduce the amount of waste going to landfill. Local authorities now require households to sort their recyclable domestic refuse (e.g. paper, plastic, aluminium tins and garden waste). The results have been substantial — whereas in 2000/2001 only 11% of domestic waste in England was recycled, in 2011 the proportion was 40%. In 2011 more commercial and industrial waste was recycled than was sent to landfill.

Further reductions in landfill will be needed in future. New technologies such as **anaerobic digestion** will help reduce organic waste (e.g. garden waste). Some councils have introduced fortnightly rather than weekly kerbside bin collections, and a few have considered charging householders for the amount of waste they produce.

> **Anaerobic digestion** is the breakdown of biodegradable material in the absence of oxygen by microbes to produce methane and carbon dioxide.

Recycling also benefits the environment by reducing GHG emissions. Recycling currently saves 18 million tonnes of carbon dioxide emissions a year in the UK.

Transport and its management

Revised

Traffic congestion

Traffic congestion imposes economic and environmental costs. There were 34 million motor vehicles in the UK in 2009 (compared with just 4 million in 1950) and the UK has some of the most congested roads in the world. Delays caused by congestion cost the country an estimated £30 billion a year. Figure 4.2 shows how average journey times by road have slowed in London in the past 30 years.

The environmental costs of traffic are also high. In Santiago, the capital of Chile, public transport is based on hundreds of old diesel buses. As well as causing congestion, these add significantly to air pollution. One effect is to raise nitrogen oxide levels to twice the WHO limit. High levels of airborne pollutants affect the health of the city's inhabitants, raising morbidity (ill health) and mortality levels. Annual costs of traffic-generated pollution exceed US$500 million. Particulates from diesel engines (three times the WHO limits) cause respiratory problems, which are a specific threat to children.

Management

Table 4.4 outlines some of the schemes that have been set up to alleviate congestion on the UK's roads.

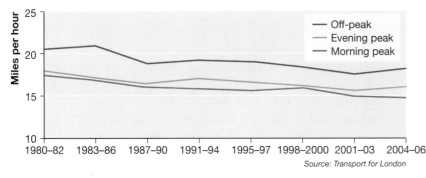

Figure 4.2 Average traffic speeds in London, 1980–2006

Table 4.4 Management responses to urban traffic congestion in the UK

Scheme	Description
Congestion charge	Introduced in London in 2003. It initially covered the central area and was extended to parts of west London in 2007. Motorists pay £8 a day to enter the congestion zone. During the first 3 years of the scheme congestion was reduced by 25%.
Light rail, trams, etc.	Several large cities such as Sheffield, Nottingham and Manchester have built light rail and tram systems in the past 20 years. These systems have proved popular with the public. They have reduced private car journeys and congestion, shortened journey times, and lowered pollution (air and noise) at street level.
Park-and-ride schemes	Large parking lots on the edge of town linked to the city centre by frequent shuttle bus services. Schemes are in place in cities such as Norwich, Oxford and Durham.
Bus lanes	Dedicated bus lanes, usually in operation during morning and evening rush hours.
Car sharing	Lanes on motorways reserved for cars with two or more occupants.
Integrated transport schemes	Integration of rapid transit (light rail, trams), bus and rail services and other forms of public transport at key interchanges. Synchronised timetables for rapid transit, bus and rail.

The Urban Transport Plan for Santiago aims to reduce traffic congestion and pollution by:

- reducing average journey lengths
- improving air quality by legislating to reduce emissions of nitrogen dioxide
- improving access to public transport
- improving mobility

It is hoped to achieve these aims by expanding the subway system and suburban rail services, creating segregated bus lanes, converting taxis to run on compressed natural gas fuel rather than diesel, reforming the public bus system, and road pricing.

Investment in public transport, and especially rapid transit, tram and light rail systems, is an important step to reducing traffic congestion and increasing sustainability in cities. In the USA, cities such as New York and San Francisco have efficient and highly successful rapid transit networks. In contrast, Los Angeles has relied heavily on its motorways and private cars. It began to invest in rapid transit only recently (Table 4.5).

Table 4.5 Rapid transit and light rail systems in Los Angeles, New York and San Francisco, 2007

City	Network (km)	Stations	Weekday passengers
Los Angeles	118	62	300,000
New York	369	468	5,000,000
San Francisco	167	43	375,000

Examiner's tip

Most students will know from experience some of the measures taken to tackle the problems of waste and traffic congestion where they live. However, convincing exam answers on these topics will require details and facts set in a specific geographical context.

Now test yourself

11 How has the CBD in large cities changed in the past 20–25 years?
12 Why is landfill an unsatisfactory method of disposing of solid waste?
13 Describe the benefits of recycling urban waste.
14 Describe the costs of traffic congestion.
15 Describe three approaches to tackling problems of urban traffic congestion.

Answers on p. 132

Check your understanding

1 Explain the link between globalisation and world cities.
2 Outline the drivers of contemporary urbanisation in cities in LEDCs.
3 Describe the main changes that have taken place in the internal distribution of population in cities in MEDCs in the past 80 years or so.
4 Examine the factors responsible for urban decline in recent decades in many MEDCs.
5 With reference to examples, describe the methods used to address problems of urban decline in MEDCs.
6 How have planners in the UK responded to the relative decline of retailing in the CBD?
7 How are cities in MEDCs becoming more environmentally sustainable?

Answers on p. 132

Exam practice

Answers and quick quiz 4 online

Section B

1. (a) Study Figure 4.1 which shows the growth and changing distribution of the global urban population 1950–2020. Describe and comment on the changing global urban population 1950–2020. [7]

   (b) With reference to examples, explain the reasons for contemporary urbanisation in the developing world. [8]

   (c) Using examples, discuss the social, economic and environmental impact of urbanisation in the developing world. [10]

2. (a) Study Table 4.3 which shows population decline in some US cities 1990–2010. Comment on the characteristics and causes of urban decline in many cities in MEDCs in the past 20 or 30 years. [7]

   (b) Explain the causes and the impact of the recent decentralisation of urban retailing and other services in MEDCs. [8]

   (c) With reference to a specific example, evaluate the impact of urban regeneration schemes. [10]

Section C

3. To what extent has urban regeneration helped to arrest social, economic and environmental decline in cities in MEDCs? [40]

4. Discuss the view that rapid contemporary urbanisation in the developing world is unsustainable. [40]

Answers and quick quiz 4 online

Online

Examiner's summary

✔ Case studies of contemporary urbanisation processes and retail change must be learned to add detail and substance to examination answers.

✔ The processes operating in large urban areas — economic growth/decline, population change — are dynamic and are driven by economic, social and technological changes.

✔ Dynamic urban processes often give rise to problems and conflicts that are resolved by managers and planners in both the public and private sectors.

✔ The causes of urban geographical changes in economic activity and population distribution can summarised as 'push' and 'pull' factors.

✔ The impact of urban decline has social and economic, as well as environmental dimensions.

✔ Urban decline can be considered at both intra-urban and inter-urban scales.

✔ The study of urban regeneration requires not only knowledge and understanding of specific schemes, but also an evaluation of their impact.

✔ To access the higher grades, technical terminology used in the study of urban areas, such as 'hierarchy', 'threshold' and 'urbanisation', must be learned and applied appropriately.

✔ Sustainability, in the context of cities, can be measured in economic and environmental terms.

5 Development and globalisation

Development and globalisation

Development and globalisation

Development   Revised ☐

The term 'development' has an economic, as well as a social and cultural meaning. **Economic development** concerns the process of creating wealth, by mobilising human and natural resources. It usually results in more jobs and higher incomes, allowing people easier access to resources such as food, energy, housing and clothing, and essential services such as education and healthcare.

But development is not just measured in economic terms. Although closely linked to income, development also refers to improvements in the **quality of life**. By this reckoning, development may mean less environmental pollution, shorter working hours and more leisure time, less crime, greater equality of opportunity for women, greater political freedom, and so on.

> **Typical mistake**
>
> Economic development does not necessarily result in an improvement in the quality of life. For instance, economic development often widens the gap between rich and poor, disrupts family life, and causes environmental degradation.

Changes associated with development

Development results in economic, demographic, social and political changes (Table 5.1).

Table 5.1 Development and change

Economic change	There is a shift in employment from agriculture and other primary industries into manufacturing and modern service activities. Productivity (i.e. the value of output per person) increases and international trade expands. As economic growth gathers pace, manufacturing moves up the value chain to focus on technologically more advanced products with higher value added. Income inequality, however, often widens.
Demographic change	Mortality rates fall as living standards improve and medical technology becomes more widely available. Fertility declines with the availability of artificial contraception, lower infant mortality and the disadvantages of large families in an industrial and post-industrial economy. With declining mortality and increasing life expectancy, populations begin to age. These changes are summarised in the demographic transition model.
Social change	Secondary education becomes widely available (especially to girls), changing values and lifestyles, and narrowing gender inequality. An ever-growing proportion of the population is urban. The marketing of western consumer products and services changes traditions (e.g. diet), values and lifestyles.
Political change	As large segments of the population experience rising living standards, education becomes universal and people are exposed to modern media (e.g. television, internet), there is pressure for improved governance and political change. Ultimately this may lead to a move towards democracy and free elections.

The development continuum

At the national scale development is often reported in polarised terms. Thus, countries are either **more economically developed (MEDCs)** or **less economically developed (LEDCs)**; first world or third world; or belong to the rich North or poor South.

However, such dichotomies are simplistic. They hide the truth that development is a continuum, with the richest and poorest countries at the extremes of the distribution (Figure 5.1). Most countries occupy the middle ground. Some, currently undergoing rapid economic development, are labelled **newly industrialising countries (NICs)** or **emerging economies**. Examples include

Exam practice answers and quick quizzes at **www.therevisionbutton.co.uk/myrevisionnotes**

China, Brazil, India and Malaysia. A clearer and more accurate classification of economic development is based on income. Using this criterion, the United Nations recognises four development categories: high income, upper middle income, lower middle income and low income countries (Table 5.2).

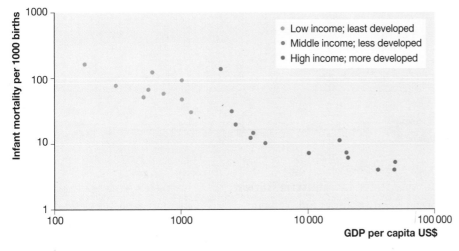

5 Development and globalisation

Examiner's tip

While the classification of countries into MEDCs and LEDCs is a useful shorthand, in essay-type questions answers must convey the idea that development is more complicated than this simple dichotomy implies.

Figure 5.1 GDP per capita and infant mortality

Table 5.2 Development indicators, 2007

Income	Life expectancy at birth (years)	Total fertility rate	Primary school completion (%)	FDI per capita (US$)	Malnutrition (% under 5s)
High income	79	1.8	99	1,527	0
Upper middle income	71	2.0	99	254	0
Lower middle income	69	2.3	91	70	25
Low income	57	4.2	65	25	28

Source: World Bank

Globalisation

Revised

Globalisation is an economic process that involves increases in the flows of ideas, people, products, services and capital between countries and on a worldwide scale. In the past 30 years or so these flows have intensified, resulting in a much greater level of integration in the world economy and closer economic ties between countries. Though its impact has been geographically uneven, globalisation has increased world production of goods and services, expanded **international trade** and brought prosperity to millions.

Globalisation has resulted in:

- a greater proportion of manufactured goods being produced by TNCs
- large increases in international trade reflecting the growth of global production and consumption
- the location of production in LEDCs, with increased flows of raw materials, manufactured goods and components from the developing to the developed world
- offshoring of services such as ICT, billing and customer care from MEDCs to LEDCs
- greater dependence on global flows of capital, controlled by major world cities such as New York, London and Tokyo
- freer international movement of goods, capital and people

Globalisation also has a cultural dimension. Companies such as McDonald's and Starbucks market their products on a global scale, transmitting American consumer values and lifestyles. People in developing countries often aspire to western consumerism, which may undermine traditional values and ultimately create more homogeneous societies.

Patterns and processes

Newly industrialising countries   Revised ▢

In the last quarter of the twentieth century the economies of South Korea, Taiwan, Hong Kong and Singapore were transformed by **industrialisation** (Table 5.3). Collectively, these newly industrialising countries (NICs) were known as the Asian 'tiger' economies. Today, several much larger economies, most notably China, Brazil and India, are being similarly transformed by industrialisation.

The key to initial economic growth in the Asian 'tigers' was the close involvement of governments. South Korea's government planned the country's economic development by:

- protecting its own industries from foreign competition (e.g. tariffs on imports)
- giving subsidies and cheap loans to export industries
- encouraging the growth of specific industrial sectors

Table 5.3 GDP per capita: Asian 'tiger' economies 1970–2009 (US$)

	1970	2009
Hong Kong	4,744	30,872
Singapore	4,308	39,423
South Korea	2,356	19,296
Taiwan	2,107	32,410

In the 1970s the emphasis was on heavy export industries such as steel, chemicals and shipbuilding. By the 1990s the Korean economy had moved up the value chain, with an increasing emphasis on higher technology sectors (e.g. electronic goods, semi-conductors, motor vehicles). At the same time large Korean companies (*chaebols*) such as Samsung, Hyundai and Kia expanded globally, investing heavily in production facilities in North America and Europe.

China   Revised ▢

In the past three decades China has experienced spectacular economic growth. This growth can be traced back to the liberal economic reforms introduced by the Chinese government from 1978 onwards. These reforms gradually shifted China from a moribund, **command economy** to a dynamic market-based one. Open economic zones such as Shanghai, Guangzhou city and the Pearl River Delta region were established to promote **free trade**, **foreign inward investment** and economic growth. Eventually these zones were extended to cover large parts of eastern China. The success of China's economic reforms and its development programme can be gauged from the fact that the private sector now generates 70% of the country's GDP. Annual economic growth, driven by exports of manufactured goods, has exceeded 10% since 2002. Today China has become the world's leading manufacturing country and by 2020 will overtake the USA as the world's largest economy. In the past decade Chinese companies have begun to invest massively overseas, particularly in Asia and Africa. Meanwhile living standards and domestic consumption (the future driver of economic growth) have risen dramatically.

> **Examiner's tip**
>
> Explanations of China's spectacular economic growth must take account of a political system that allows government to implement reforms in the interests of the state, regardless of opposition.

Globalisation of services

Revised

For decades TNCs have located manufacturing production overseas to take advantage of lower costs and proximity to foreign markets. Since the 1990s, thanks to advances in telecoms, a range of service activities have also been **offshored** from Europe and North America to low-cost economies such as India.

India has been a favoured destination because of its large surplus of English-speaking graduates and wage rates that are only a fraction of those in the developed world. Initially call centres for utility companies, internet service providers, internet banking and so on were outsourced. TNCs such as HSBC, British Airways, American Express and Tesco also transferred 'back office' jobs, which don't require direct contact with clients, overseas. By 2008, India had 4 million service jobs offshored by North American and European companies, worth $US57 billion per year. By Indian standards these jobs are well paid: they also provide much-needed employment for women.

But offshoring of services is no longer confined to low-level and 'back office' activities. Increasingly, highly skilled operations such as research, computerised design, software design, product development and marketing are being moved offshore. ICT services now employ 1.5 million in India and account for 7% of the country's GDP. Bangalore in southern India has attracted massive foreign investment and has become India's ICT capital. It has state-of-the-art business parks, modern innovation centres and direct flights to New York, London and Tokyo.

> **Examiner's tip**
>
> India's development as a global service centre demonstrates the power of telecoms technology and the decreasing significance of distance in the location of economic activity.

Growth in the twenty-first century

Revised

In the course of the next three decades the influence of the world's largest emerging economies — the so-called BRICS of Brazil, Russia, India, China and South Africa — will continue to grow. By mid-century they could surpass the established economic powers of the USA, Japan and the EU in the global economy.

The BRICS potential for growth stems from their favourable demographics, large economies and abundant natural resources (e.g. oil and gas in Russia, mineral ores in Brazil and South Africa). Already the BRICS account for 40% of the world's population and one-quarter of global GDP. Economic growth rates among the BRICS countries averaged 6.5% between 2000 and 2010 — nearly two and a half times the global average. So by 2050, Goldman Sachs predict that China, India, Brazil and Russia will have become the world's first, third, fifth and sixth largest economies, respectively.

Future growth of the BRICS is due in part to their large youthful populations and to their emerging middle classes (1.6 billion by 2020). Favourable demographics, combined with the growing prosperity of the educated middle class, will drive up demand and so generate huge markets for businesses.

Technology will also play a role in this new economic order. Investment in science research and development is already at record levels in China and by 2020 will amount to 2.5% of the country's GDP. ICT and innovation will also have a critical bearing on business, government and individuals, and will operate as engines of growth. In China these developments have helped to foster the growth of high technology industries such as aerospace, pharmaceuticals, computers and medical equipment.

> **Examiner's tip**
>
> Descriptions of China's economy are problematic. China is already an economic superpower and the world's largest manufacturing nation. However, in other respects China lags well behind MEDCs. GNI per capita was US$5,370 in 2010; according to the World Bank this makes China an 'upper middle income' country.

Countries at low levels of economic development

The world's least developed countries are mainly concentrated in sub-Saharan Africa (SSA). According to the UN's Human Development Index (HDI) for 2010, the world's ten poorest countries were in this region; and of the poorest 20 countries only one, Afghanistan, was located outside SSA. The World Bank defines the world's poorest countries as 'low income countries' with gross national income (GNI) per capita less than US$995 per year. In 2009, 43 out of 213 countries were classed as 'low income' and 28 were in SSA. These countries are characterised by:

● poor nutrition, poor health, and inadequate education and adult literacy levels

● economic vulnerability, with dependence on primary production (especially traditional agriculture), trade instability and indebtedness

Quality of life
Revised

The UN defines **quality of life** as the 'notion of human welfare (wellbeing) measured by social indicators'. In countries with low levels of development, poverty is widespread. In 2006, 1 billion people survived on less than US$1 per day (the World Bank's definition of extreme poverty); another 1.7 billion people were classified as poor, struggling to live on just US$1–2 per day.

Income directly affects quality of life, though the latter is difficult to measure. Among the factors that influence quality of life are environment quality (e.g. clean drinking water, improved sanitation), security (e.g. crime rates), political stability, vulnerability to natural hazards and so on. For example, in rural Afghanistan barely 40% of the population has access to safe drinking water, and less than one-third to improved sanitation. The earthquake that struck Haiti in 2010 is estimated to have killed between 250,000 and 300,000 people and destroyed 250,000 houses, underlining the vulnerability of the poorest nations to natural hazards.

Political instability and poor governance is also prominent in many of the world's poorest countries. In 2011, 15 countries in Africa (e.g. Congo DR, the Sudan, Somalia) were involved in wars. The outcome is thousands of refugees, food insecurity, disruption to family life and disease. Corruption among government officials and political leaders is widespread. Money that should be used for development (including foreign aid) is often channelled to the military and the governing elite. The recent history of Zimbabwe underlines the importance of good governance. Once the 'bread basket' of southern Africa, food production collapsed between 2000 and 2005, leaving nearly half the population malnourished and reliant on food imports and food aid. The cause was entirely political. Government land seizures destroyed the country's highly efficient farming industry, hitting food exports, **food security** and rural incomes.

Debt
Revised

Compared to their incomes, many low income countries carry huge debt burdens. Loans from foreign banks, governments, the IMF and the World Bank must be paid with interest. These debts can be crippling. Interest payments divert much-needed capital for development to rich countries in North America, Europe and east Asia. Between 1980 and 2005 the world's poorest countries made debt repayments of US$550 billion on loans worth US$540 billion. Yet at the end of this period they still owed US$523 billion! Despite repeated promises by the wealthiest G8 countries to cancel debts, few have been implemented.

In low income countries the total foreign debt in 2010 was US$108 billion (31% of GDI), equivalent to US$177 per person. However, in some countries the ratio of debt to GDI was much higher (e.g. Liberia 515%, Congo DR 118%). In Liberia the average foreign debt per person was US$918, while annual GDI per person was just US$160.

Social problems
Revised

Social problems in low income countries include population change, ill health, education and literacy, and gender inequality (Table 5.4).

Table 5.4 Social indicators of development in some low income countries

	Life expectancy (years)	Annual population growth rate (%)	Total fertility	Child mortality 0–5 years (per 1,000)	Adult female literacy rates (%)	HIV/AIDS prevalence (%) (adults 15–49)
Congo DR	48	2.6	7.1	125	62	4.2
Niger	52.5	3.7	6.9	177	15	0.8
Mozambique	48.4	2.1	4.6	198	64	12.5
Chad	48	2.0	5.1	162	23	3.5

Population change

In many of the world's poorest countries population growth (e.g. 2% or more per year) outstrips economic growth, contributing to poverty and high rates of dependency. There are exceptions — in SSA, HIV/AIDS has raised mortality and will slow population growth for at least 20 years. Indeed, South Africa's population is even forecast to decline in the period 2011–2030. This too has implications for development, reducing the proportion of economically active adults to the rest of the population.

Ill health

SSA is not only the world's least developed region, it also has the highest incidence of HIV/AIDS. In Swaziland one-third of young women and one-fifth of young men are HIV positive. Because HIV/AIDS is most severe among younger adults, it has devastating social effects. Children are orphaned (11.6 million in Africa alone), families destroyed and there are fewer adults to care for aged relatives. The economic effects are even worse. At the national level health services are overburdened, key skilled workers in healthcare and education are lost

> **Examiner's tip**
>
> It should be noted that a stagnant or declining population creates economic problems just as serious as those resulting from rapid population growth.

and chronic illness makes workers less productive. On an individual household level HIV/AIDS causes a loss of income, increased expenditure on medicines, indebtedness and a greater likelihood of poverty.

Education and literacy

Education and literacy are key ingredients of **human capital**. In most poor countries, low levels of education and literacy are a consequence as well as a cause of lack of development. The UN estimates that around 600 million women in LEDCs are illiterate and that more than 100 million receive no formal education at all. In Africa's poorest countries literacy rates among women are less than 25%, and in Niger the figure is just 15%. Average years of education for men and women in some of the least developed countries are shown in Table 5.5.

Gender inequality

In many LEDCs religious practices and cultural values discriminate against women and neutralise a large part of the potential workforce. Education of girls is often given lower priority than for boys (Table 5.5). As a result, productivity is reduced and development is held back. **Gender inequality** is most pronounced in SSA, south and southwest Asia and Arab states. SSA's poorest countries have the highest gender inequality, although the most extreme inequality is found in two Islamic states: Afghanistan and Yemen.

Table 5.5 Average years of education

	Women	Men
Afghanistan	5	11
Central African Republic	5	8
Chad	5	9
Eritrea	4	6
Mozambique	7	9
Pakistan	6	8

Typical mistake

The causes of poverty in countries in LEDCs are complex, and involve the interaction of many variables. No single variable is of overriding importance. Discussions of poverty in LEDCs must show an appreciation of the multivariate nature of the problem.

Global social and economic groupings

North and South   Revised

At a global scale, the terms rich North and poor South provide a convenient summary of global development. The so-called **north–south divide** is delimited by the **Brandt Line**, named after Germany's chancellor in the 1980s. However, this geographical dichotomy is only a broad generalisation. In fact many of the world's poorest countries, especially in Africa and Central America, are in the Northern Hemisphere, while rich countries such as Australia and New Zealand, and middle income countries such as Chile and Argentina are in the south. This reminds us that development is a continuum, with countries in many different stages in the development process.

Groupings of nations: the EU   Revised

Many countries combine to form regional **trade blocs** such as the European Union (EU) and the North American Free Trade Agreement (NAFTA) (Figure 5.2). With an internal market of 500 million people and 27 member states, the EU is the largest as well as the most integrated of regional trade blocs. The EU promotes the interests of its member states by:

- removing internal barriers to trade (e.g. tariffs), capital and people between its members
- creating a common currency (adopted by 17 EU states), which eases the movement of goods, services and people across internal borders

- protecting its own industries from foreign imports by imposing a **common external tariff** and **quotas**
- subsidising the exports of selected industries

Agriculture is the most heavily protected industry in the EU. In addition to tariffs on foreign food imports some agricultural exports are made artificially competitive in foreign markets by **subsidies**.

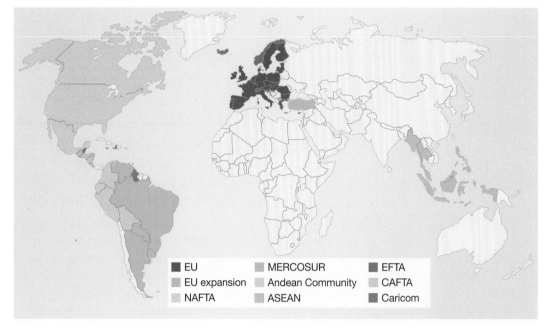

Figure 5.2 World trading blocs

Consequences of national groupings

National groupings such as the EU form powerful economic (and political) blocs. The EU gives member states exclusive access to the huge European market and, for small countries such as Belgium and Portugal, greater influence when competing with the economic superpowers of the USA, China and Japan. While the EU serves the economic interests of member states its impact is felt well beyond Europe.

- The EU and NAFTA control 53% of world merchandise exports.
- EU subsidies to agricultural exports undermine the competitiveness of some farmers in the developing world, exacerbating unemployment and poverty.
- EU tariffs exclude many non-EU farm products (especially those of small, high cost producers in the developing world) from the EU market.
- Difficulties faced by traditional trading partners outside the EU in accessing the EU market (e.g. New Zealand, Australia). Special bilateral trade agreements have been made between some EU member states and their traditional trading partners (e.g. banana growers in the Caribbean and the Pacific and importing countries such as France and the UK).

The World Trade Organisation (WTO) deals with the rules of international trade and seeks to promote free trade. A major on-going dispute between the WTO and the EU concerns the liberalisation of international trade in farm products, ending export subsidies, and the opening-up of the EU market to third-party producers.

Examiner's tip

You should be aware of the debate concerning the overall impact of free trade. While the removal of trade barriers has caused an expansion of global trade and fuelled globalisation, weak economies in LEDCs, exposed to open competition from economic superpowers and regional trade blocs, could suffer economic collapse.

Now test yourself

6 Why are the world's poorest countries concentrated in sub-Saharan Africa?

7 Construct a table that lists and describes the social problems typical of the world's poorest countries.

8 What is meant by the term 'human capital'?

9 In what ways do poor countries fail to develop their human capital?

10 What are the advantages of trade blocs, such as the EU, to member states?

Answers on p. 133

Aspects of globalisation

Transnational corporations

Transnational corporations (TNCs) are large enterprises with production and/or service operations in several countries and which compete in global markets. They are the driving force behind the globalisation of industry, services and trade of the past 50 years. TNCs have complex and diverse organisations and structures. Some are highly centralised with decision-making and strategies concentrated at a single headquarter location (e.g. IKEA). Others, like Honda, are more devolved, with regional organisation geared to the needs of local markets (see Table 5.6).

Case study) TNCs: IKEA and Honda

IKEA
Background IKEA is a Swedish TNC and the world's largest furniture retailer. IKEA was founded in 1943 in the small town of Älmhult in southern Sweden. The company employs 118,000 workers worldwide and operates stores in 35 different countries.

Structure Company HQ is in Leiden, near Amsterdam, the hub of Ikea's management and control. Amsterdam is a major urban centre with excellent communication and transport links.

Production and research & development (R&D) Most production is outsourced. Subcontractors (1,350) are located worldwide in 50 countries. Supplies are delivered to distribution centres and retail stores on a 'just-in-time basis', minimising capital tied up in stock. IKEA has its own production arm — Swedwood — employing 13,000 people, mainly in Poland. Älmhult in southern Sweden is the main design centre. Design is also outsourced, but mainly within Sweden.

Honda
Background Established in 1948, Honda is a Japanese TNC. Its core activity is automobile manufacture. It is Japan's second largest automobile maker (after Toyota) and the world's fifth largest. In 2010, Honda's automobile sector had a US$106 billion turnover. 85% of Honda's revenue is from automobile manufacture. However,

Honda is a diversified company. It is the world's biggest manufacturer of motorcycles and engines (automobile, marine, jets). It also makes robots, garden equipment and provides financial services. Honda employs 180,000 people worldwide.

Structure Honda's global HQ is in Tokyo but it has regional HQs in all its main markets. This structure allows Honda to serve regional markets more efficiently and respond to geographical differences in culture, environmental protection laws and regional development.

Production and research & development (R&D) Honda is the most globalised TNC in the automobile industry. Its first overseas plant was located in Ohio in 1982. Subsequently Honda located car assembly plants in the USA, Canada, UK, Brazil, Argentina, Indonesia, Thailand, China, India and Pakistan. These plants give direct access to regional markets and therefore avoid trade barriers. Major production sites are Saitama and Suzuka in Japan, Ohio and Alabama (USA), Alliston (Canada), Swindon (UK), and Ayutthaya (Thailand). Four out of five Honda cars are made outside Japan, 40% in North America. North America is Honda's biggest market. R&D is located in Japan and in regional markets such as North America and Europe. The market for cars in MEDCs is mature and Honda will look increasingly to the BRICS and developing countries for future growth.

Car-making giants such as Ford and General Motors have adopted a devolved geographical structure, with North American, European and other divisions. Within these regional divisions, the companies source most of their parts from independent suppliers and manufacture models that reflect regional tastes and preferences. Ford and General Motors also produce models for the global market. This allows them to achieve **economies of scale**, making fewer models but in greater numbers.

The dominance of TNCs in the global economy is summarised as follows:

- TNCs are responsible for four-fifths of global economic output
- the top 500 TNCs account for 90% of **foreign direct investment** (FDI)
- TNCs generate two-thirds of world trade; one-third of world trade is intra-firm trade between TNCs

Exam practice answers and quick quizzes at **www.therevisionbutton.co.uk/myrevisionnotes**

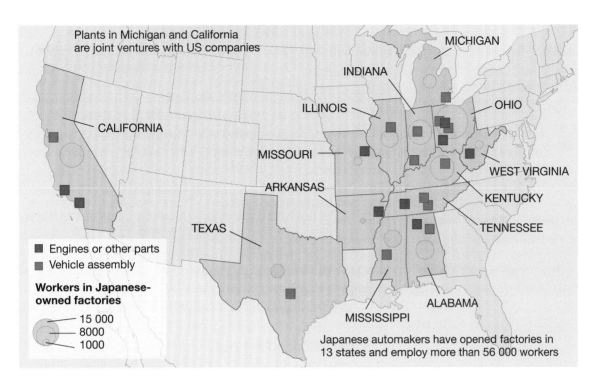

Plants in Michigan and California are joint ventures with US companies

MICHIGAN
INDIANA
ILLINOIS
OHIO
CALIFORNIA
MISSOURI
WEST VIRGINIA
ARKANSAS
KENTUCKY
TEXAS
TENNESSEE
ALABAMA
MISSISSIPPI

■ Engines or other parts
■ Vehicle assembly

Workers in Japanese-owned factories
— 15 000
— 8000
— 1000

Japanese automakers have opened factories in 13 states and employ more than 56 000 workers

Figure 5.3 Japanese and Korean automotive investment in the USA

Reasons for the growth of TNCs

Revised

The emergence of TNCs as the dominant players in the global economy is due to:

- their flexibility
- their size and political influence
- the growth of global telecoms

Flexibility

TNCs exploit geographical differences in the cost of **factors of production** (e.g. natural resources, labour, capital) and state policies towards **inward investment**. Most governments want to attract FDI and, in competition with other countries, offer incentives such as reduced taxes and subsidies. However, the relationship between states and major TNCs is not always equal. Given their flexibility, TNCs can easily switch production elsewhere to lower-cost locations or to countries where financial inducements are more attractive. For example, in 2006, Peugeot, the French TNC, located production of the new 207 model in the Czech Republic rather than at its existing Coventry plant. The motive was economic — labour costs in the Czech Republic were only one-fifth of those in the UK. The result — closure of the Coventry factory and the loss of 2,300 jobs.

Size and political influence

Major enterprises such as TNCs have access to world markets. This favours large-scale production of goods and services, allowing them to achieve economies of scale, lower unit costs and out-compete rivals. Because of their economic power (providing employment, extra domestic spending and exports), governments are eager to attract TNCs. This places TNCs in a strong position to negotiate favourable conditions such as financial subsidies and tax concessions. The advantage of overseas locations for TNCs is the access they give to foreign domestic markets (whether individual countries or trade blocs) which otherwise would be restricted by tariff barriers and quotas.

Typical mistake

TNCs vary enormously in size, function, organisation and the extent to which they are truly globalised. It is important to recognise that TNCs are not just confined to the manufacturing sector. Many well-known companies in the service sector, such as McDonald's, Starbucks, Goldman Sachs and HBSC, are TNCs.

Global telecoms

Advances in telecoms have been a driving force behind globalisation and the expansion of TNCs. In the past decade, the success of global retailers such as Amazon has been based entirely on the internet and e-marketing. Global telecoms networks allow TNCs to access and share information across their geographically dispersed operations, coordinate and implement management strategies, and organise production and sales in diverse cultural and political environments.

> **Typical mistake**
>
> Globalisation is not a new process. Ford Motors (US TNC) established its first plant in the UK in 1911. But because of modern telecoms, the pace of globalisation has accelerated in the past 20–30 years. Telecoms have greatly assisted the globalisation of TNCs in the service sectors (e.g. producer services such as law, accounting, banking, advertising).

The impacts of TNCs   Revised

Social and economic issues in developed countries

TNCs have the power to shape the global economy. Because they control flows of capital, materials, components and technical expertise — all of which are crucial to economic development — individual governments are not inclined to challenge them. However, although FDI has economic advantages, it also creates disadvantages (Table 5.6).

Table 5.6 Social and economic impacts of TNCs and FDI on national economies

Advantages	Disadvantages
Provide inward investment and create jobs for local people.	Many jobs are relatively low-skilled in labour-intensive industries (electronics, clothing). In developed countries, capital-intensive, foreign enterprises may offer few higher paid managerial, development, design and marketing job opportunities.
Increase incomes and raise living standards among employees.	Lack of security, as TNCs switch operations to lower-cost locations elsewhere (see Hoover-Candy case study). Decisions are driven by company budgets, not the social and economic interests of local communities.
Boost exports and help the trade balance.	Lack of control, with key investment decisions taken overseas at company headquarters.
Develop and improve skill levels and expertise among the workforce, and technology and process systems among local firms.	TNCs may demand further government financial incentives not to disinvest.
Increase spending and create a multiplier effect within local economies.	Competition can lead to the closure of domestic firms.
Attract related investment by suppliers and create clusters of economic activity.	

Case study   Closure of Hoover-Candy's factory at Merthyr Tydfil, 2009

Background Hoover-Candy is an Italian TNC with its HQ near Milan. The company employs 7,610 people worldwide, 6,000 outside Italy. It has 38 branch plants. It manufactured washing machines and tumble dryers at its Merthyr Tydfil plant. The plant was originally opened by the US TNC, Hoover, in 1948. Merthyr Tydfil was an early example of FDI and at its peak it employed more than 5,000 workers.

Closure Hoover-Candy closed its Merthyr Tydfil plant in March 2009. The reason given was its high operating costs, making it uncompetitive with plants in China and eastern Europe. Production was transferred to a low-cost location — Turkey.

Impact 450 people worked at the plant. Although some operations remained on the site (e.g. warehousing, after-sales), closure of the plant resulted in 337 job losses. Hoover-Candy's demise marked the end of large-scale manufacturing in Merthyr Tydfil. The largest employers are now the council, the health service, a T-Mobile call centre, the Welsh Assembly and Tesco. Unemployment has risen rapidly — up to 6.5% by the end of 2008 — with 16.5% of the workforce on incapacity benefit.

The lack of employment has led to a fall in the town's 55,000 population. Merthyr is the only district of Wales where further depopulation is expected in future. By almost any measure — social wellbeing, health, educational attainment, wages, life expectancy, alcohol abuse, house prices — the town is struggling to survive.

Environmental impacts

It is claimed that one reason why TNCs **offshore** manufacturing to LEDCs is to exploit weak environmental protection laws and their lax enforcement. As a result, costs are passed to the environment as pollution, and to local people who suffer ill health and disease.

Since the 1960s the US–Mexico border region has proved hugely attractive to FDI from US, European and Asian TNCs. A wide range of goods is manufactured on the Mexican side of the border, including clothes, chemicals and electronics. Location in the border region provides low labour costs and proximity to US markets.

But large concentrations of industry along the border create environmental problems. Air pollution from hundreds of factories is responsible for respiratory disease, cardiovascular disease and premature death. The worst pollution is in the twin cities of El Paso (USA) and Ciudad Juarez (Mexico). Ciudad Juarez has more than 300 foreign-owned factories. Air pollution from particulates, ozone, carbon monoxide and nitrogen oxide often exceeds Mexico's air quality standards and pollution drifts across the border and affects El Paso.

However, not all of the pollution originates from factories connected to foreign TNCs. Blame also rests with the pace of globalisation and economic growth and the absence of enforcement of air quality standards in Mexico.

Examiner's tip

Large TNCs are controversial, bringing economic and social benefits and disbenefits to communities. You should be aware of the pros and cons of inward investment by TNCs and be prepared to argue a point of view, supported by evidence and examples.

Typical mistake

Levels of pollution that are unacceptable in MEDCs are tolerated in many LEDCs. Pollution emitted by local factories often exceeds that from TNCs operations (see Mexico–US border example left).

Development issues

Trade versus aid
Revised

Two contrasting approaches to development aim to (a) improve terms of international trade for LEDCs, and (b) increase the flow of development aid to LEDCs.

Unfair trade

The WTO supports free trade, arguing that it ultimately increases the volume of international trade to the benefit of all countries. But international trade often works against the interests of the poorest countries. Small farmers in the developing world growing export crops such as coffee and bananas have relatively high unit costs and struggle to compete with large-scale **agribusiness** enterprises based in MEDCs. In addition, the prices farmers get for their crops are often only a tiny fraction of the wholesale or retail price.

Developing countries also complain that many rich countries subsidise their agricultural exports, thus violating the principle of fairness in trade. For example in 2008 the WTO declared the US government's US$3 billion subsidy to its cotton growers illegal. The result was a major victory for poor farmers in West African countries such as Burkina Faso and Senegal who rely on cotton exports as their only source of cash. In future these farmers should get 5–12% more for their cotton, giving household incomes a significant boost and helping to lift thousands of families out of poverty.

Fair trade

Fair trade is an informal movement initiated by businesses and consumers in MEDCs to give poor farmers and cooperatives in LEDCs a greater share of the value of their food exports. Because consumers pay a higher price for fair trade products, farmers' cooperatives in LEDCs receive extra money, which can be spent on individual families or community projects. In the past 10 years or so, fair trade products such as coffee (e.g. Cafédirect) and chocolate (e.g. Divine) have been successfully marketed in developed countries. Nearly 100 fair trade products are now available in the UK and in 2007–2008 global sales of fair trade tea and coffee doubled.

Development through trade

Expansion of trade in LEDCs should, in theory, stimulate economic activity, provide increased incomes, spending, jobs and tax revenue. If growth is sustainable it should provide long-term economic benefits, including improvements in living standards and the quality of life. However, there are caveats. First, most poor countries rely on a narrow range of **primary goods** exports (e.g. food, minerals) whose prices fluctuate on world commodity markets. And second, as free trade expands, businesses in the developing world may struggle to compete with powerful TNCs backed by huge resources of capital and technology.

> **Examiner's tip**
>
> Successful analysis of development issues should invoke a balanced discussion of ideas and values that often conflict. You should reach a personal viewpoint that is consistent with your discussion, and the weight you have given to each argument.

Case study   EU and banana imports

Background Close political and historical ties between the UK (and other EU countries) and small banana-producing nations in Africa, the Caribbean and Pacific (ACP) gave rise to preferential trade agreements between the EU and these countries. Without preferential treatment, small, labour-intensive enterprises would be unable to compete with highly efficient US TNCs operating large plantations in Central and South America (e.g. Dole, Chiquita). In this instance the trade agreement provided a guaranteed market for ACP growers but effectively discriminated mainly against US banana TNCs. It also increased the cost of bananas to EU consumers.

Dispute The preferential trade agreement between the EU and the ACP caused a long-running dispute between the EU and the WTO. Non-ACP bananas paid a tariff of US$250 per tonne on entry to the EU market while ACP growers enjoyed a 775,000 tonne tariff-free banana quota. The WTO in 2007 ruled that the agreement between the EU and ACP violated global trade rules and gave ACP growers an unfair advantage. The dispute was essentially a battle between powerful corporate business, and small farmers whose existence depended on the preferential trade agreement with the EU.

Outcome An agreement was reached in 2009. The EU will end its preferential treatment for ACP growers and slash its tariffs on non-ACP banana imports by 2017. At the same time the EU will compensate ACP producers, paying them US$300 million to help meet competition and lower their costs.

International aid

International aid comprises money and resources transferred from rich countries and **multilateral agencies** (e.g. UN, IMF) to LEDCs. Crucially, donor countries and organisations do not expect full or direct repayment for aid. There are two types of development aid — **bilateral aid**, from a donor government to a recipient country and **multilateral aid**, given by donor agencies such as the UN and the World Bank. In 1970 the UN set an aid target for individual countries of 0.7% of GNI. Just five donor countries achieved this figure in 2010. In 2009 the UK government spent £7.4 billion on international aid (0.52% of GNI), two-thirds of which was bilateral.

Development through international aid

International aid is a means of improving the quality of life of people in the world's poorest countries and achieving the UN's eight **millennium goals** set out in 2000. But international aid's effectiveness as a catalyst for development and improving the lives of the poor has a mixed record.

● In the past both bilateral and multilateral aid has funded wasteful infrastructure projects (e.g. dam construction) with little direct benefit for the poor and at the same time having adverse environmental effects.

- Aid has often been diverted to military spending, extravagant construction projects such as presidential palaces and parliamentary buildings.
- Widespread corruption and inefficient governance have often meant that international aid fails to reach its intended target (i.e. the poor).
- Some recipient countries receive aid even though they are capable of financing their own development. For example, in 2009 India received £334 million of aid from the UK — more than any other country.
- In the past, bilateral aid has often been 'tied' to the donor country, which stipulated how the money could be spent. This usually meant using resources, equipment and technical assistance from the donor country, rather than from the lowest-cost provider. Arguably, 'tied' aid benefits the donor country as much as the recipient.

New approaches

Today, the emphasis has shifted to small-scale development aid projects funded by **non-governmental organisations (NGOs)**, government departments and multilateral development agencies. They are seen as a more effective way of directing aid to those most in need. This 'bottom up' approach starts at the grass-roots level. It supports involvement by local people and uses their skills and labour, encouraging them to take ownership as **stakeholders**. Typical small-scale aid projects include improving irrigation and farming techniques, reafforestation programmes, sinking village wells to provide clean drinking water, improving sanitation systems, building schools and microfinance schemes. Aid at this scale, aimed at small communities and which is clearly targeted and sustainable, can effect a real and lasting improvement in the lives of thousands of poor people in the developing world (see case study on northern Nigeria).

> **Non-governmental organisations (NGOs)** are non-profit-making, voluntary groups that operate independently of government. They pursue social aims such as improving the quality of life of the poor. Examples include Oxfam, Water Aid, and the Red Cross.

Case study Girls' Education Project, northern Nigeria

Background The Girls' Education Project (GEP) was established in 2005. Its aim is encourage more girls in Nigeria's six northern (predominantly Muslim) states into primary education (8 million children of primary school age in Nigeria — mostly girls — are not in education). In northern Nigeria the ratio of boys to girls in primary education is 3:1. The project is consistent with the Millennium Development Goals to achieve universal primary education for all children, promote gender equality and empower women. GEP is a joint project between the Nigerian government, UNICEF and the UK Department for International Development. Funding is £50 million, mainly from the UK. It currently applies to just 10% of primary schools in the region.

Scheme The project's success hinges on getting more women teachers into rural primary schools. Scholarships are provided for trainee teachers who will return to their own villages to teach. Women teachers will act as mentors and role models for young girls. In a Muslim society, parents are more likely to send their daughters to school if they are taught by women. Free education materials are distributed to pupils in 720 schools, helping parents send their children to school.

Outcomes Between 2005 and 2008 there was a 15% improvement in primary school enrolment in northern Nigeria. Although it is too early to assess the impact of GEP, it is expected that when rural schools start to receive a steady flow of local qualified women teachers hundreds of thousands of primary students will benefit. Education is also an essential tool in providing human capital for future development in rural Nigeria.

Economic versus environmental sustainability Revised

Rapid and sustained economic growth in NICs during the past 30 or 40 years has often been at the expense of the environment. This reflects the priority given to economic expansion, job creation and raising living standards, rather than environmental protection.

China

Recent industrialisation and urbanisation in Shanxi province and the Pearl River Delta (see case study) have resulted in severe pollution and environmental degradation. **Acid rain**, caused mainly by sulphur dioxide emissions from coal-fired power stations, has caused major environmental and health problems in southeast China. Meanwhile, Beijing, Shanghai and most other large cities in China suffer chronic air pollution (especially **particulates**) with pollution levels two to three times higher than London and Los Angeles. At the global scale, China is the world's largest emitter of carbon dioxide and other greenhouse gases thought to be responsible for global warming and climate change.

However, with the advance of economic development, China is beginning to address environmental issues. Acid rain control is a high priority and **desulphurisation** technology is being applied to China's coal-fired power stations. By 2030 China plans to generate 20% of its energy from **renewables**. Already China leads the world in investment in wind power and aims to expand production by 100 GW between 2010 and 2015. The Three Gorges Dam, completed in 2009, generates 18,000 MW of clean energy, and four new HEP schemes costing US$60 billion were announced in 2011. There is also significant interest in solar and geothermal energy.

Case study Economic development and environmental pollution in the Pearl River Delta, China

Background Guangdong in southeast China has been at the centre of China's drive to industrialisation and economic growth (Figure 5.4). The province has a population of 95 million in an area comparable in size to England and Wales. At the industrial heart of Guangdong is the Pearl River Delta (PRD) Special Economic Zone, which is geared to the export of manufactured goods to North America and western Europe.

Industrial development Globalisation has been the driver for industrial development in the PRD and Guangdong. The Chinese government encouraged development by making the region a free-trade zone. This, together with its coastal location, has attracted major inward investment. Hong Kong is the main investor, followed by Taiwan, Japan, South Korea, Singapore and the USA. Many of the world's leading TNCs, including Hitachi, Samsung, Sony, Siemens, Pepsi and Toyota, have located production in the PRD. This has helped to make the province one of the most prosperous in China. In 2006 the PRD's GDP per capita was 2.5 times greater than the provincial average. Prolonged economic boom (evidence of economic sustainability) has

made significant reductions in poverty. Around four-fifths of Guangdong's GDP derives from the PRD.

Environmental problems Industrial development has stimulated massive in-migration (from the countryside) and urbanisation. In the PRD, 77% of the population lives in towns and cities. However, treatment of sewage and industrial effluent has not kept pace with urban and economic growth. Three-quarters of Guangdong's cities have no sewage treatment plants, polluting rivers and threatening water supplies. The Pearl River is grossly polluted and downstream from Guangzhou it is virtually lifeless.

Air pollution is also a major concern. Sulphur dioxide from coal-fired power stations and manufacturing industries (the main cause of acid rain) often rises above safe levels in Foshan and Guangzhou, especially in winter. Pollution from nitrogen oxide from motor vehicles, ozone and particulates shows a similar pattern (Figure 5.4). Guangzhou is affected by brown smog from motor vehicles 130 days a year on average.

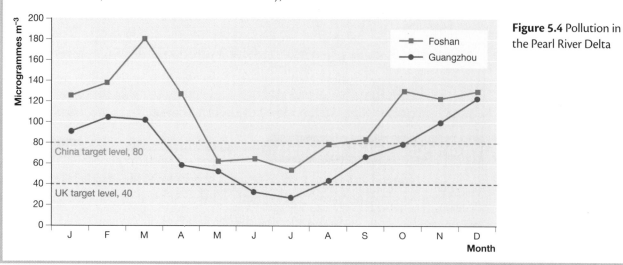

Figure 5.4 Pollution in the Pearl River Delta

Exam practice answers and quick quizzes at **www.therevisionbutton.co.uk/myrevisionnotes**

Brazil

Brazil is the world's fifth largest country by area and by population. Like China, Brazil is an upper middle income country, which in the past 30 years has experienced spectacular industrialisation and economic growth. Between 1960 and 2009 the value of Brazil's manufacturing sector increased by more than 50 times, and output of hydroelectricity (HEP) grew nearly ninefold. GDP per capita rose by 50% between 2000 and 2009. Meanwhile, Brazil's population soared from 54 million in 1950 to 201 million in 2011.

Brazil's economic growth is sustainable and will continue in future. It is a vast country with huge natural wealth in minerals (iron ore, manganese, tin, bauxite and others), HEP, oil and timber. Its large population, which is highly urbanised (87% urban), will provide the human capital needed to power future development. Most of Brazil remains sparsely settled. Forty per cent of the population is concentrated in the southeast, within 100–200 km of the coast.

Brazil's rapid economic development and massive population growth have created huge environmental pressures that have resulted in deforestation, land degradation, pollution and loss of biodiversity (Table 5.7).

> **Typical mistake**
>
> While environmental degradation has often accompanied industrialisation in China, recent efforts by the Chinese government to achieve a more sustainable economic growth are having some success and will gather momentum in future.

> **Examiner's tip**
>
> Study of NICs, emerging economies and industrialisation in MEDCs in the past, suggest that economic development imposes significant environmental costs. You should consider whether economic development need be at the expense of the environment and whether environmental degradation is a cost worth paying for development.

Table 5.7 Environmental impact of development in Brazil

Deforestation	The area of farmland in Brazil increased from 6.6 million ha in 1920 to 64 million ha in 2003. Farmland has replaced natural ecosystems (especially forest) causing loss of biodiversity. Deforestation continues today, though at slower rates than in the past half century. Even so, the forest area still declined by 7% in the period 1990–2010. The loss of 150,000 km² between 2000 and 2006 (mostly rainforest in Amazonia) was equal to an area the size of Greece. 65–70% of deforestation is caused by cattle ranching. Large-scale commercial farming, mainly for soybeans, accounts for up to 25%. Other losses are due to logging, urbanisation, fires and road building. Until the mid-1990s, the government granted legal title to farmers settling in the Amazon rainforest. Up to150,000 small farmers took up the challenge. However, permanent cultivation of rainforest soils proved unsustainable. Most plots were eventually abandoned and converted to low-grade pasture for ranching.
HEP	Several massive HEP projects have been completed, providing electricity for the cities and industries of the Atlantic seaboard. The environmental impact of dam construction includes deforestation, loss of natural ecosystems, increased GHG emissions and the creation of anoxic conditions in dam reservoirs (caused by decaying vegetation). Little consideration was given to the environmental impact of huge dams such as the Tucurui Dam on the Tocatins River in Para state. Currently, four major dams are under construction — Xingu, Madeira, Tapajos and Tocatins — which will add 25,000 MW to Brazil's electricity capacity.
Air pollution	Polluted air in Brazil's two mega cities — São Paulo and Rio de Janeiro — from traffic, power stations, heavy industry and domestic fires, are a health risk. The pollutants include PM_{10} particulates, NO_2, SO_2 and O_3. Interaction of O_3 with sunlight produces photochemical smog and levels of pollution often exceed WHO guidelines.

Sustainable tourism: myth or reality?   Revised ☐

Types of tourism

Many governments in the developing world promote international tourism, rather than industrialisation, as their preferred vehicle for economic development. They market natural resources such as tropical climates, exotic landscapes and biodiversity to attract wealthy tourists from North America, Europe and Japan. This type of tourism, promoted in countries such as Kenya, Costa Rica and Nepal, is known as **ecotourism**. Unlike **mass tourism**, ecotourism aims to minimise its impact on natural environments and ecosystems while creating economic opportunities for local people.

> **Mass tourism** caters for large numbers of visitors, concentrated in small geographical areas and is based around natural resources such as beaches and mountains. It invariably has a damaging impact on the environment.

Ecotourism overlaps with **sustainable tourism**. Sustainable tourism is an attractive concept for environmentally conscious tourists because it claims to do no permanent damage to natural resources, economies and cultures. It is, however, debateable whether a truly sustainable tourism can ever exist. Even in the most strictly controlled conservation areas, such as the USA's national parks, the pressures of tourism often degrade fragile habitats and wildlife. If sustainable tourism is achievable, visitor numbers must remain small. Yet for tourism to make a difference to economic development it has to attract large numbers of visitors. The temptation for governments is to expand tourism, attract ever larger numbers of tourists and increase tourism revenues, all of which run counter to the ideals of sustainable tourism.

Costa Rica — sustainable tourism? Revised

1980s and 1990s — sustainable growth

Costa Rica is a small country in Central America with a population of 4.6 million (2011). In the 1980s Costa Rica was one of the first countries to specialise in ecotourism, based on its exotic mix of natural environments — volcanoes, plateaux, coast, cloud forest — and rich biodiversity. The government created 24 national parks that, together with Indian reserves, covered 28% of the country (Figure 5.5). Areas along the Pacific and Caribbean coasts were also protected as marine reserves. The government encouraged ecotourism by awarding certificates to businesses that practised sustainable tourism, and blue flags to beaches that met safety and water quality standards. By the 1990s tourism had become Costa Rica's leading economic sector.

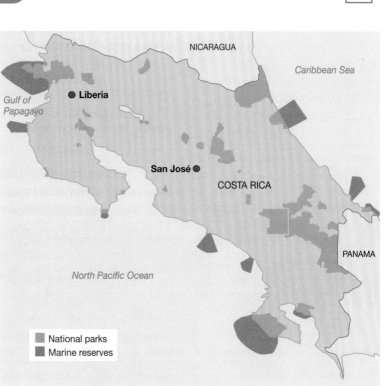

Figure 5.5 Costa Rica: national parks and protected marine areas

During the 1990s tourism adhered to the sustainable model. Hotels were small and largely owned by native people, visitor numbers were relatively modest, and there was little damage to the natural environment and local cultures. Costa Rica gained a reputation as a world leader in environmentally friendly tourism.

2000–2011 — the intrusion of mass tourism

The first decade of the twenty-first century brought sweeping changes to Costa Rica's tourism industry. These changes were related to tourism's unprecedented growth. By 2011 tourism contributed 13% of the nation's GDP and indirectly generated 250,000 jobs. In 2008 visitor numbers topped 2 million for the first time (half from the USA) and the viability of sustainable tourism began to be questioned.

This boom was largely due to large-scale beach resort and holiday home developments on the Pacific coast. These developments were in direct conflict with Costa Rica's reputation for ecotourism and an undamaged environment. Only limited regulation was applied to the building of small hotels, vacation homes, apartment towers and gated communities along the coast. The government encouraged the construction of all-inclusive resorts that included hotels, golf courses, restaurants, spas and shops. It also created the infrastructure to attract investment for a large-scale tourism project centred on the Gulf of Papagayo. The main stimulus to growth was the opening of the international airport at Liberia in 2002 and direct flights from the USA. In the 2000s Costa Rica's tourism industry seemed increasingly drawn towards the economic rewards offered by the mass market.

Environmental impact

In the past decade tourism development in Costa Rica and the pursuit of profit has been at the expense of the sustainable use of natural resources, especially in the fragile coastal environment (Table 5.8).

Table 5.8 Environmental impact of tourism development on Costa Rica's Pacific coast 2000–2010

Water	Demand for water has risen steeply (a golf course requires as much water as a small town of 5,000 people). Water shortages occur during the dry season, when there is insufficient to meet the needs of local communities. Overexploitation of aquifers has led to a fall in the water table. Aqueducts have been constructed to transfer water to the coastal zone.
Pollution	Some prime tourist beaches have been polluted by discharges of raw sewage into the coastal environment and rivers by hotels, resorts, businesses and vacation homes. In 2005, nearly three-quarters of Pacific beaches were threatened by pollution.
Forests and mangroves	Despite legal protection, forests and mangroves have been lost to unrestricted tourism development. The scale and pace of development has often meant that environmental protection laws have not been implemented by local authorities.
Scenic beauty	Unregulated development (often on prime sites) has had a negative impact on the scenic qualities of the coastal zone. By law the public has free access to all beaches but new development has, in places, prevented de facto access.

Conclusion

Costa Rica set out in the 1980s to establish a tourism industry based on the sustainable use of the country's natural resources. To a large extent this was achieved in the first decade of the industry's development. However, rising economic expectations and the lure of profits attracted foreign investment and a move towards a type of mass tourism in the early twenty-first century. This was driven by a number of factors, including direct flights from the USA, proximity to North America, political stability and the country's well educated workforce. This new tourism is no longer environmentally sustainable.

> **Examiner's tip**
>
> With rising economic expectations can governments initially committed to small-scale ecotourism resist the pressures for mass tourism? You need to consider this and other issues concerning tourism in your revision programme.

Now test yourself

11 What are the main features of TNCs?

12 How have TNCs influenced the globalisation of the world economy?

13 Why is it difficult for many LEDCs to achieve economic development by expanding their international trade?

14 What is the difference between bilateral and multilateral aid?

15 Why has international aid often failed to improve the lives of the poorest people in LEDCs?

16 What is meant by the terms ecotourism, sustainable tourism and mass tourism?

Answers on p. 133

Check your understanding

1 Examine the contention that economic development is a continuum.

2 Outline the main drivers responsible for the globalisation of manufacturing and service industries.

3 Consider the role of TNCs in the globalisation of the international economy.

4 What are the major obstacles to development in the world's poorest countries?

5 Explain how it is possible to promote effective economic development in poor countries through international trade and aid.

6 Under what circumstances is it possible for governments to deliver sustainable tourism?

Answers on p. 134

Exam practice

Section B

1 (a) Study Figure 5.1 which shows the relationship between GDP per capita and infant mortality in low, middle and high income countries. Comment on the relationship between GDP per capita and infant mortality. **[7]**

(b) Explain the recent industrial development of China. **[8]**

(c) With reference to specific examples, discuss the impact of economic development on the environment. **[10]**

2 (a) Study Figure 5.6 which shows the global distribution of GDP per capita. Comment on the global distribution of GDP per capita in Figure 5.6. **[7]**

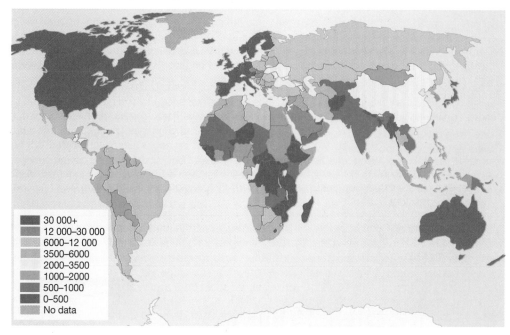

30 000+
12 000–30 000
6000–12 000
3500–6000
2000–3500
1000–2000
500–1000
0–500
No data

Figure 5.6 Global GDP per capita, 2007

(b) With reference to a TNC you have studied, describe and explain its main characteristics and its spatial distribution. **[8]**

(c) With reference to examples from contrasting areas of the world, discuss the view that sustainable tourism is a myth. **[10]**

Section C

3 'The operation of TNCs in the global economy creates social, economic and environmental advantages and disadvantages.' Discuss with reference to one TNC you have studied. **[40]**

4 'Globalisation increases social and economic inequality.' To what extent do you agree with this view? **[40]**

Answers and quick quiz 5 online

Online

Examiner's summary

✔ Be aware that economic development has a downside — it does not benefit everyone (it often increases inequality) and damages the environment.

✔ The division of countries into MEDCs and LEDCs is a gross simplification. Many countries belong to a middle income group between the two extremes. Discussions should reflect this complexity.

✔ The internet and telecoms are major drivers of today's unprecedented globalisation.

✔ The causes of poverty in LEDCs are complex and result from the interaction of many variables. Discussion must convey the multivariate nature of poverty.

✔ Free trade has helped to stimulate globalisation but has not benefited all nations, particularly the poorest.

✔ TNCs are important in the globalisation of service activities as well as manufacturing. Modern telecoms have greatly increased the scope for globalising services.

✔ TNCs are controversial and raise important issues. Arguments for both sides must be revised and viewpoints supported by actual examples.

✔ Discussions of issues must be balanced and must demonstrate how a chosen viewpoint outweighs alternative arguments.

6 Contemporary conflicts and challenges

The geographical basis of conflict

The origin and nature of conflict is grounded in a range of socio-economic, political, cultural, historic, resource, territorial and ideological factors. Rarely do these factors operate in isolation.

Identity

Revised

Conflict between groups occurs at local, regional and national scales. **Localism** is where issues and conflicts arise at a small scale, such as a neighbourhood or village, and a commonality of place unites people. Pressure groups formed to oppose development on greenbelt land, the siting of wind turbines, the closure of a local post office or hospital are examples of localism.

Regionalism refers to a larger geographical area and unlike localism involves a common sense of history, geography and culture. A number of areas in England, such as Cornwall and the northeast, have strong regional identities. Cornwall's identity stems from is relative isolation from the rest of England and its geography as a peninsula, its outlook on the sea, its Celtic traditions, its distinctive accent and dialect and, in the past, its own language. All of these factors give Cornish people a distinct identity and separateness from the rest of England.

Nationalism takes identity a stage further. A nation is a large group of people who share some of the following characteristics: the same language, history and culture and a common ethnicity or race. If a nation has its own sovereign government it is termed a **nation state**. However, within sovereign states there are often several nations. In Spain, for instance, the Catalans and the Basques have a strong sense of national identity. Conflicts occur when nations generate separatist movements and demand their own sovereign government. Many recent conflicts have their origins in national identities and the desire for independence. Examples include the Basques in Spain, the Tamils in Sri Lanka, the Kurds in Iraq, the Chechens in Russia and many others. Many civil wars in Africa stem from nationalism in the form of inter-tribal conflicts.

> **Examiner's tip**
>
> Localism, regionalism and nationalism cover the geographical spectrum of conflict. The seriousness of conflict increases with geographical scale, i.e. the larger the scale, the more likely that conflict will result in violence.

Territory and resources

Revised

Territory

At the heart of many conflicts between nations are disputes about territory. The on-going conflict between Israel and the Palestinians centres on territorial claims to the land of Israel and Jerusalem. After 60 years of conflict between Israel and her Arab neighbours, no solution has been reached. The Falklands War in 1982, between the UK and Argentina, developed from Argentinian claims of sovereignty over the South Atlantic islands. Meanwhile the dispute between India and Pakistan over the territory of Kashmir is unresolved. Territorial conflicts are often settled by compromise between protagonists through **partition**. This was the outcome of twentieth-century conflicts in Ireland, Sudan and the Korean peninsula.

Resources

Decades of civil war in Sudan, which killed an estimated 2.5 million people, ended in 2011 when South Sudan became an independent state. The conflict was as much about controlling South Sudan's natural resource wealth (oil, gas, diamonds, gold, timber) as it was about cultural and ethnic differences between the Islamic north, and the Christian south. However, formal agreement between Khartoum and Juba on dividing oil revenues (the oilfields are located along the new border, 75% in South Sudan) has still to be reached.

In many parts of the world, economic development and population growth have increased the demand for water, while climate change has reduced water supplies. The result has been international disputes over water resources that risk spilling over into war. Conflicts arise where rivers such as the Nile and Euphrates drain a number of states. Both rivers have their headwaters in humid regions, and flow to the coast across arid and semi-arid regions. Egypt, a largely desert state, relies almost entirely on the River Nile for its water. However, plans by Ethiopia to build a huge dam (5.2 MW) on the Blue Nile near the border with Sudan threaten Egypt's water supplies. As a result both countries could be drawn into conflict.

Syria is already in conflict with Turkey over water rights on the Euphrates River. Turkey's East Anatolia Project involves building a series of dams on the river, reducing flows downstream and increasing salinity levels. Syria argues that the effect on its agriculture would be disastrous.

> **Typical mistake**
>
> Natural resources are often an unstated aim of armed conflict and wars. Many observers believe that recent wars between the Allies and Iraq were as much about oil as they were about the military threat presented by Iraq, or regime change on ideological grounds.

Ideology   Revised ☐

Ideology is a theory or set of beliefs or principles on which a political system is based and is a powerful driver of international conflict. The clash between **democracy** and **communism** is the most obvious example of recent ideological conflict. Between 1945 and the early 1990s it led to a 'cold war' between democratic countries led by the USA, and the communist governments of the Soviet Union and eastern Europe. The cold war produced an armed stand-off in Europe, as well as an **arms race** and the stockpiling of huge arsenals of nuclear weapons. However, a number of proxy wars between communism and democratic powers were fought outside the European theatre. They included the Vietnam war of the 1960s and the Soviet invasion of Afghanistan in the late 1970s. The collapse of communism in the eastern bloc ended the cold war and the ideological rivalry between the USA and Russia.

Today communist ideologies survive in China, and in a handful of small states such as North Korea and Cuba. But the power of ideology to provoke conflict remains. Al-Qaeda, a loose-knit international terrorist organisation united by a common ideology, was responsible for many atrocities such as the 11 September 2001 (9/11) attacks in New York, the 7 July 2005 (7/7) bombings in London and the Madrid train attacks in 2004. Al-Qaeda preached a radical version of Islam, with terror used as a weapon in pursuit of the ultimate **geopolitical** goals of world domination and the establishment of an Islamic caliphate.

> **Geopolitics** is the effect of geography on international politics, e.g. location of resources, access to the ocean, size/extent/shape of a territory, situation of a country in relation to others, etc.

Patterns and expression of conflicts

Revised

Conflicts occur at national, regional and local scales. They may be violent or non-violent and expressed through war, terrorism, political activity and debate.

National conflicts

The most serious conflicts are violent confrontations or wars between states. Fortunately most international conflicts are not serious enough to result in war. Conflicts relating to trade, transnational boundary issues such as acid rain, immigration, fishing quotas and so on are normally settled peacefully by negotiation and diplomacy.

Because war is costly and its outcome unpredictable it is usually a last resort. In 2011 the NATO allies, led by France and the UK and supported by the UN, took control of all military operations against Libya. The immediate aim was humanitarian — to protect the Libyan anti-government rebels from annihilation by Gaddafi's armed forces. The longer-term aim was ideological — to support the **insurrection**, topple the Gaddafi dictatorship and democratise Libya. Even the limited warfare in Libya was costly. Apart from injury and loss of life, the economic costs were huge. The first two months of war cost the British government £220 million. High-tech military equipment is not cheap — a single cruise missile costs £500,000, and refuelling a strike aircraft such as a Tornado jet costs £30,000!

Regional conflicts

The most serious regional conflicts result in **civil war**. Recent civil wars have been fought in Sudan, Somalia, DR Congo, Rwanda and Sri Lanka. These conflicts often polarise around ethnic and racial differences, and the perception by minority groups of oppression by the majority. This scenario often creates **separatist** movements.

Local conflicts

Local conflicts affect limited areas, and are usually confined to small communities and self-interest groups. The route of the proposed High Speed Two (HS2) rail line between London and Birmingham has aroused localised but vociferous opposition. Nowhere has the opposition been stronger than in the Chilterns, a densely wooded chalk upland and Area of Outstanding Natural Beauty, to the northwest of London. The area is home to many wealthy and influential residents who have conducted an effective and concerted campaign against HS2. General objections relate to the overall cost of the project (up to £30 billion), as well as its environmental impact (landscape, wildlife, noise). More specific objections come from residents who may see the value of their homes decreasing. These self-interest groups are often labelled as **NIMBYS**.

Supporters of the HS2 include central and local government, and business leaders in the Midlands and the North. They argue that a high speed rail network, connecting the Midlands (and eventually northern England) to London and the continent, will stimulate economic growth in the regions and reduce spatial inequality in the UK economy. Ultimately this battle will be expressed through petition, demonstrations, lobbying of parliament, discussion and debate.

Examiner's tip

The pattern and expression of any conflict is often dependent on geographical scale. Questions asking for discussion of general statements relating to conflicts could assess their validity by using the concept of scale.

Typical mistake

It should be remembered that rarely are there simple answers to local conflicts. However, although arguments may be finely balanced, ultimately a value judgement in favour of one party, whether by an individual, a group of experts or politicians, has to be made.

Conflict resolution

Possible resolutions to conflict include:

- negotiations and/or a legal treaty that formally recognises an end to conflict and specifies terms and conditions
- partition, e.g. Sudan, Korea, Ireland
- the military defeat of one party, e.g. Tamil Tigers in the Sri Lankan civil war, 2009
- the internal collapse of a government and its political structures, e.g. Tunisia and Egypt, 2011
- the policing of a region by a neutral third-party, multilateral force, e.g. UN, Ivory Coast and Liberia
- decisions made by central and local government, and quangos (e.g. Primary Care Trusts) following informed debate on local issues (e.g. wind farms, green belt violations, hospital closures). These decisions are often subject to appeal and even to independent inquiries. They may also be influenced by elections, lobbying of MPs, and parliamentary debate
- devolution of power to regional bodies, e.g. creation of the Scottish Parliament and the Welsh Assembly
- the creation of an overarching international body such as the UN, able to function as a mediator in disputes and dedicated to peace; and international organisations such as the EU, where strong economic ties and common policies reduce the likelihood of armed conflict between member states

Conflict over the use of local resources

Foveran Dunes versus golf development

Background

The Sands of Forvie, which lie 20 km north of Aberdeen in northeast Scotland, is the UK's fifth largest coastal dune complex. It comprises fixed and mobile dunes, sand beaches, a sand **spit** at the mouth of the Ythan estuary, dune pastures, marshes and heaths. Owing to its spectacular landforms, rich habitats and ecology, the Sands of Forvie has **National Nature Reserve** (NNR) status. At the northern end of the dune system the Foveran links are protected as a **Site of Special Scientific Interest** (SSSI) (Figure 6.1).

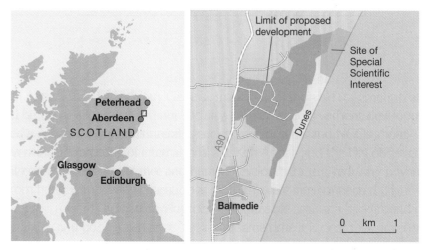

Figure 6.1 Foveran dunes

Despite the protected status of the Foveran links, in June 2008 Donald Trump, a US property billionaire, applied to Aberdeenshire County Council for permission to develop a £750 million golf resort at Foveran. The proposed development consisted of two 18-hole championship golf courses, 950 timeshare flats, a 450-bed hotel and 500 holiday homes.

Conflict and conflict resolution

Trump's plan for development proved controversial. Conservationists were outraged at a scheme that would destroy a significant part of the Foveran dunes. Local residents were also strongly opposed for social and environmental reasons. In November 2007 local councillors rejected the plan. However, the Scottish government unexpectedly 'called-in' the plan and launched a **public enquiry** into the scheme.

Public enquiry

Opposition to the proposal came from a number of environmental agencies, including the Scottish Wildlife Trust, the RSPB, the WWF and the Botanical Society of the British Isles. They argued that development would destroy a significant part of the Foveran dunes, supposedly protected by its SSSI status. Local residents set up their own protest group called *Tripping up Trump*. Several believed they would lose their homes and farms if the development went ahead. Others feared the development would change fundamentally the character of the area and its agricultural and fishing heritage.

At the public enquiry the development received the backing of Scotland's first minister, local businesses and tourism agencies. The Trump organisation submitted evidence on the positive economic impact of the scheme. They estimated that around 2,000 full-time jobs would be created locally, which with **multiplier effects** would amount to nearly 6,000 jobs across Scotland. Overall the scheme would inject around £40 million a year into the local economy. Advocates of the scheme said that conservationists had exaggerated the environmental impact which would only affect one-tenth of the area occupied by the Foveran dunes.

Outcome

The four-month-long public enquiry ended in October 2008, and a month later the Scottish government approved the scheme, believing the economic benefits outweighed the environmental costs. However, strict conditions were imposed on the developers to minimise environmental damage. Several local residents refused to sell their homes and farms to the Trump organisation, despite threats of compulsory purchase from the Scottish government.

Work on the new golf courses started in July 2010. Plans to build a hotel and hundreds of homes still awaited the approval of local councillors in summer 2011. The course should be completed by July 2012 and will be officially opened by Sir Sean Connery and Scotland's first minister, Alex Salmond.

Examiner's tip

You should note that conservation status such as SSSI is no guarantee that development in a conservation area will not take place.

Examiner's tip

Discussion of issues requires students to formulate their own view on a controversial subject. This will involve weighing the conflicting arguments and making a decision that reflects the student's own values and priorities.

Now test yourself

1 What is meant by the terms localism, regionalism and nationalism?
2 What is the difference between a nation and a state?
3 What is an ideology? Give examples.
4 List the ways in which issues that result in conflict can be resolved.
5 What is separatism? Outline its possible causes.
6 How and why has the construction of a new golf course at Foveran created conflict?

Answers on p. 134

The geographical impact of the Arab–Israeli conflict

Background to the conflict

The Arab–Israeli conflict is rooted in competing claims for territory and resources. It has its origins in the late nineteenth century when the Zionist movement campaigned for the Jews to have a single homeland.

Two thousand years ago Israel was the Jewish homeland. In the first century AD a Jewish revolt against the Roman Empire was crushed, Jerusalem and its temple destroyed, and the Jews dispersed throughout the Mediterranean and the Middle East. This scattering of the Jewish people is called the **diaspora**. For the Zionists there was one problem — the territory of Israel was not vacant; it had been part of the Ottoman Empire and occupied by Palestinian Muslims for hundreds of years.

Table 6.1 Timeline: Arab–Israeli conflict

1917	Balfour declaration. Britain 'viewed with favour the establishment of a national home for the Jews in Palestine'.
1919–1921	Zionist organisation submits plans for implementation of the Balfour declaration. Arab demonstrations and riots in Jerusalem against Jewish immigration.
1923	British mandate for Palestine, approved by UN, comes into operation.
1929–1935	Sporadic acts of violence by Islamist groups against Jews and British in Palestine.
1936–1939	Arab revolt. 5,000 Arabs and 400 Jews killed.
1937–1939	Attempts to achieve a diplomatic solution (including partition) fail. Jewish immigration continues.
1940–1947	Radical paramilitary Jewish groups active (e.g. Irgun). King David hotel in Jerusalem, the centre of British administration in Palestine, bombed in 1946, killing 91 people.
1947	UN plan to divide Palestine into separate Jewish and Arab states is rejected by Arabs (Figure 6.2).
1948	Israel declares independence of British rule. The first Arab–Israeli war lasts 13 months. Lebanon, Syria, Iraq, Egypt and Jordan in conflict with Israel.
1949	700,000 Arabs become refugees as a result of the war. Many Jews flee to Israel from Arab states.
1956	Israel invades Sinai to stop infiltration of paramilitaries from Egypt and to reopen the Suez Canal (with the approval of Britain and France). Israel withdraws from Sinai the following year.
1964	The Palestine Liberation Army (PLO) is founded.
1967	The Six Day war between Israel and its Arab neighbours. Israel captures Sinai, Gaza (both from Egypt), the West Bank (from Jordan) and the Golan Heights (from Syria) (Figure 6.2). 15,000–25,000 war casualties. The captured territories give Israel more defensible borders.
1968–1972	Various atrocities committed by Arab paramilitaries, including the Munich Olympic Games massacre in 1972.
1973	The Yom Kippur War. Israel is attacked by surrounding Arab states.
1974–1978	Sporadic skirmishes and killing of civilians on both sides.
1978	Camp David accord. Israel withdraws from Sinai and agrees to enter negotiations over the West Bank and Gaza in exchange for peace with Egypt.
1979	Egypt recognises Israel's right to exist — the first Arab nation to do so.
1982	Israel invades southern Lebanon to protect its northern border. It withdraws the following year.
1987	The first Palestinian uprising or intifada in Israel — violence, strikes, riots and civil disobedience.
1990s	First suicide bombings by Hamas and other militant Arab groups in Israel. Hundreds of civilians killed.

2000	Camp David Summit collapses. Israeli leaders refuse to accept proposals that would cede most of the West Bank and Gaza to the Palestinians. Second intifada begins.
2002	Israel begins the construction of its security barrier (or wall) to separate physically the West Bank from Israel and reduce attacks by paramilitaries.
2005	Israel disengages from Gaza, which is controlled and administered by the Palestinians.
2006	Rocket attacks from Gaza on Israel are met by counter-offensives from Israel. Lebanon (Hezbollah) war. Conflict lasts for a month. 1,200 killed, mainly civilians. 300,000 people displaced.

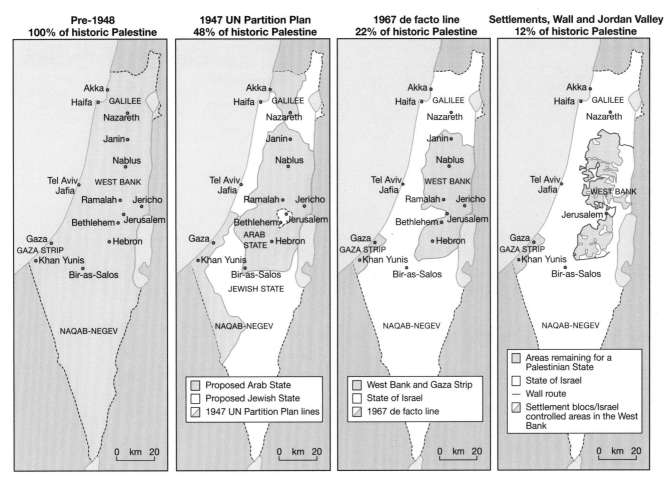

Figure 6.2 Changing boundaries of Israel and Palestine

In 1917 the British government backed the Zionist cause and called for the establishment of a Jewish state in Israel. This put the Jews on a collision course with the Palestinians, sparking a conflict that continued throughout the twentieth century and remains unresolved today (Table 6.1). Major armed conflicts between Israel and its Arab neighbours broke out in:

- 1948, when Israel declared itself an independent Jewish state, triggering the first Arab–Israeli war lasting 13 months
- 1967 (Six Day War), 1973 (Yom Kippur War) and 2006 (Lebanon War)

Between the wars there were internal uprisings in Israel and the Palestinian territories (intifada), suicide bombings, rocket attacks and aerial bombardments. It is estimated that the conflict has, in total, claimed nearly 9,500 civilian lives, more than 80% of them Palestinian. Numerous peace initiatives instigated by the USA and its allies and the UN have failed to find a lasting solution the Palestinian problem.

Examiner's tip

Although detailed knowledge of the history of the Arab–Israeli conflict will not be required in the exam, some historical background is needed to provide a context for understanding the social, economic and environmental issues in Israel and the Palestine territories.

Social issues

Thousands of people — mainly Palestinians — have been displaced by events in Israel in the past 60 years. Large numbers of **refugees** live temporarily, often in the most appalling conditions, in camps in Jordan, Lebanon, and Egypt. Gaza is a particular problem. There, 1.6 million people are crammed into an area of just 360 km². Overcrowding has created desperate living conditions — **infant mortality** is twice as high as in Israel, there are shortages of medicines and medical equipment, and disruption to electricity, water supply and sanitation services. Conditions were made worse by Israeli bombings in 2008 and the ensuing blockade on imports and exports. In 2009 the WHO reported that diarrhoeal disease and TB were on the increase and the UN warned of an impending humanitarian crisis.

Table 6.2 Economic inequalities between Israelis and Palestinians (US$)

	Israel (US$)	West Bank	Gaza
GNI per capita	18,850	1,230	
Unemployment	7.6	18.6	34.8
Poverty rate	21.6	67	80

Economic issues

Inequality. The Arab–Israeli conflict has widened economic inequalities between Israelis and Palestinians (Table 6.2). Unemployment, poverty and GNI per capita in the Palestinian territories are similar to those in low income countries in the developing world. In Gaza, opportunities for employment are limited by Israel's economic blockade and its bomb-damaged infrastructure. Even if the Palestinian territories were to become an independent state, its economic viability is compromised by the West Bank's landlocked situation and the geographical isolation of Gaza.

Jewish settlements on the West Bank. Israel's establishment of Jewish settlements on the West Bank is an unresolved issue and the cause of resentment. Currently, Israel retains full control in 60% of the area of the West Bank (4% of the total population) which includes all of the Jewish settlements. Unless Israel cedes most of this land, the West Bank will not be economically viable as an independent state.

Security wall. To counter the threat of Palestinian terrorist attacks, in the early 2000s Israel began constructing a 750 km long, 3.5 m high security wall. When complete it will physically separate Israel from the West Bank. Although the wall has improved security for Israeli civilians, it has seriously disrupted the lives of people living along the border. Palestinians can no longer commute to work in Israel and Palestinian businesses cannot easily access Israeli markets. Reduced trade is a feature of the border towns, as the barrier has cut across hinterlands and market areas. Elsewhere the barrier has split farms, reduced access to farmland, destroyed arable land, orchards and greenhouses, and disrupted irrigation networks.

Environmental issues

Water resources. Israel, Gaza, the West Bank and the surrounding Arab states is a water-stressed region. Low rainfall and high rates of evapotranspiration mean that renewable water supplies barely meet demand (Figure 6.3). Total renewable water resources in Israel and the Palestinian territories are around 1,800 million m³/year. However, water demand exceeds 2,400 million m³/year. The shortfall is made up by desalination plants and the unsustainable extraction of groundwater (Figure 6.3). Israel relies on the Jordan River catchment for 60% of its water (both surface flow and groundwater). Groundwater is also available from the Mediterranean aquifer along the coast. The Sea of Galilee is Israel's main freshwater store.

> **Examiner's tip**
>
> The social, economic and environmental issues in Israel and Palestinian territories must not be studied in isolation. Instead they must always be linked back to the political conflict that created them.

> **Typical mistake**
>
> Aridity is not just determined by low rainfall. Evapotranspiration must also be taken into account.

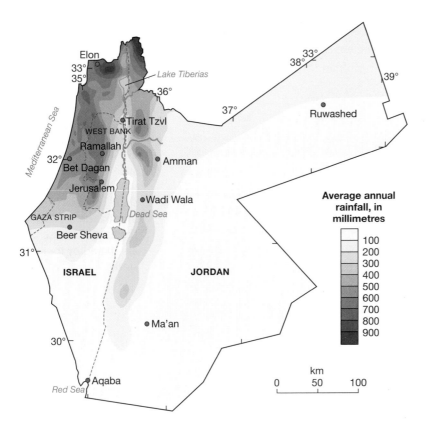

Figure 6.3 Mean annual rainfall in Israel and Jordan

Political conflict and the water crisis. Israel's water crisis has exacerbated political conflict in the region. The creation of the state of Israel led to a massive immigration of Jews from North Africa, Russia, the USA and Europe. This immigration boosted Israel's population from 2.14 million in 1960, to 7.47 million in 2011, and greatly increased the pressure on water resources. High rates of natural increase and large influxes of refugees have also produced spectacular population growth in the Palestinian territories, which now support more than 4 million people. Economic development and water-intensive agriculture have also increased water demand.

Water resources in Israel and the Palestine territories are shared unequally. Currently 80% go to Israel and 20% to the Palestine territories. Water shortage is a major constraint to economic development in the West Bank — a situation maintained by on-going conflict between the two nations. Political instability in this region has also resulted in little transnational cooperation between Israel, Syria and Jordan in integrating water resource management.

Environmental impact. Unsustainable extraction of groundwater has led to falling water tables and increases in water salinity. Both have detrimental environmental effects with springs and wetlands drying up and wildlife habitats being degraded. Five out of Israel's six indigenous amphibians are endangered and the painted frog and European water vole have recently become extinct (Figure 6.4).

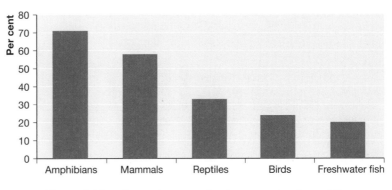

Figure 6.4 Israel's vertebrate species that are under threat (%)

The Jordan River catchment has been badly affected. A large part of the flow of the Upper Jordan River is diverted by Israel's National Water Carrier (NWC) upstream from the Sea of Galilee. Average annual water flow of the Upper Jordan River near the Sea of Galilee is 500 million m³ but average outflow is only 72 million m³/year. Freshwater ecosystems in the Sea of Galilee are threatened by declining water levels, overfishing, pollution from farm runoff and the excessive growth of phytoplankton due to nutrient enrichment.

Flows on the River Yarmouk, a major tributary of the Jordan River, have been reduced by dams in Jordan and Syria that divert water for agriculture. As a result, the Lower Jordan has only a fraction of its natural flow and its waters are highly saline. Water quality is further degraded because many West Bank settlements pour untreated sewage into the Jordan. With the Jordan River reduced to a trickle, water levels in the Dead Sea are falling by an average 1.3 m/year. The survival of the Dead Sea — the lowest elevation on the land surface, the world's deepest hypersaline lake and a potential World Heritage Site — with its unique freshwater and saline habitats and biodiversity, is at risk.

> **Examiner's tip**
>
> In a water-stressed region such as the Middle East, water has major political significance. So, too, do international rivers such as the Jordan, whose waters are shared by four different states.

The challenge of multicultural societies in the UK

The UK is often described as a **multicultural society**. Although numerically the population is dominated by people of white British descent, the UK comprises a mosaic of ethnic minority groups and cultures. In 2009, the Office of National Statistics estimated that just over 17% of England's population were 'non-white British' — a proportion that is expected to double by 2050.

The development of multicultural societies   Revised ☐

The UK's development as a multicultural society is primarily related to immigration patterns over the past 50 or 60 years.

Immigration patterns

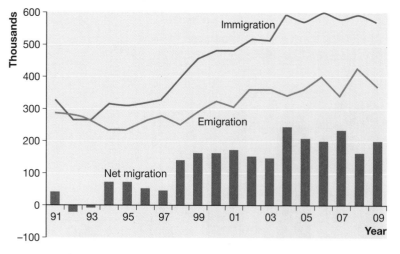

Figure 6.5 Migration and the UK, 1991–2009

International **immigration** is a long-established feature of UK demography. Large-scale immigration from Ireland occurred from the mid-nineteenth century onwards, together with waves of Jewish immigrants from eastern Europe in the second half of the century.

Since the 1960s international immigration, particularly from countries in the New Commonwealth, has occurred steadily. In the 2000s it accelerated to record levels (Figure 6.5). Between 2003 and 2009 **net migration** was in excess of 600,000, twice the number in the 1990s. Almost half of the UK's population growth in the decade 2001–2010 was due to immigration. Natural increase was also relatively high in this period, largely because of the youthfulness of immigrants.

Causes of immigration

Immigration to the UK occurs for economic, social and humanitarian reasons.

- Immigration to the UK from the Caribbean and South Asia in the period 1950–1980 was initially for economic reasons. Immigrants were often single men seeking employment in industries that had labour shortages and were unattractive to British workers. As immigrants settled permanently, legislation was gradually amended to allow marriage partners to join spouses in the UK, and later dependents and those with 'substantial connections with UK citizens'. Until the 1980s, New Commonwealth citizens were accorded privileged status to enter to the UK.

- Following the enlargement of the EU in 2004, the UK government adopted an open access policy towards immigration from new EU states (most EU countries placed limits on immigration). Contrary to government forecasts, a flood of immigrants arrived from eastern and central Europe, especially from Poland, Slovakia and the Baltic states. These economic migrants were often well educated and skilled. They were attracted by the UK's high wage economy; the prospect of accumulating capital (or remitting it home); and better job prospects than in central and eastern Europe. In total an estimated 1.9 million immigrants came to the UK from eastern and central Europe between 2004 and 2010. Not all of these immigrants settled permanently, many returned home with the onset of recession in 2008.

- In the 2000–2009 decade many immigrants entered the UK as refugees and **asylum seekers**, fleeing war and political persecution in countries such as Iraq, Afghanistan, Somalia and the DR Congo (Figure 6.6). In 2009–2010 the Home Office received more than 42,000 applications for asylum.

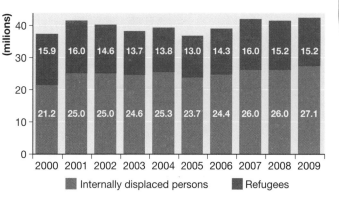

Figure 6.6 Global numbers of refugees and displaced people, 2000–2009

Other factors contributing to a multicultural society

The multicultural nature of British society is also the result of a tradition of liberalism and tolerance towards ethnic minorities. In the past 30 years, successive governments have supported **multiculturalism**. This is the idea that several different cultures (rather than one national culture) can co-exist peacefully and equitably in a country. Thus integration of ethnic minorities with the majority British culture had little priority. Today some observers believe that multiculturalism has contributed to the 'isolated and parallel lives' lived by many ethnic minority communities in the UK.

In other respects, the evolution of a multicultural society can be seen as an inevitable outcome of globalisation. Most countries in the western world are more multicultural than they were 50 years ago. Changes in transport technology, information exchange through telecoms, and the formation of supranational organisations such as the EU, have made international population movements much easier.

The geographical distribution of cultural groupings

Revised

Geographical patterns

Most ethnic minority groups have an uneven geographical distribution in the UK. South Asian and Afro-Caribbean minorities are concentrated in urban areas, especially London and major metropolitan cities such as Birmingham, Manchester, Leicester, Leeds and Glasgow (Figure 6.7). Chinese ethnic groups, however, have a more dispersed spatial pattern.

More recent immigrants from central and eastern Europe also have a dispersed pattern, inhabiting small towns and rural districts as well as metropolitan areas. For example there are large concentrations of central and eastern Europeans in the east of England and in coastal towns. Poles are the largest and most ubiquitous group, accounting for nearly two-thirds of all central and eastern European immigrants. There are unusually large concentrations of immigrants from central and eastern Europe in places as widely scattered as Boston (Lincs), Northampton, Herefordshire and Orkney.

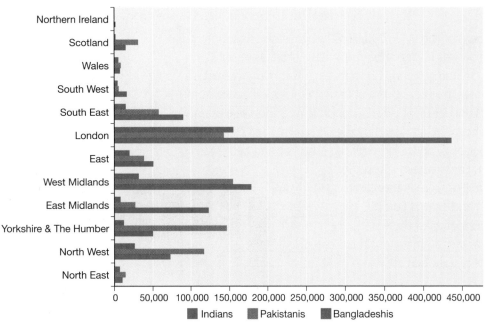

Figure 6.7 Regional distribution of Asian communities in the UK

Factors influencing the geography of ethnic minorities

At the national scale, the distribution of South Asian and Afro-Caribbean minority groups is the result of historical, economic and social factors.

- The first immigrants settled in large towns and cities where employment was available, e.g. textile industries of West Yorkshire and the East Midlands, public transport in London.

- These initial clusters attracted other immigrants with similar ethnic backgrounds. Large concentrations of Indian or Afro-Caribbean minorities provided social advantages, such as family relationships, opportunities for religious and other group activities, access to ethnic retailers and work within ethnic enclaves.

At the regional scale, ethnic minorities originating from South Asia, the Caribbean and Africa are overwhelmingly concentrated in inner and central city areas. This pattern reflects both self-segregation (e.g. access to low-cost rented housing, social and cultural advantages) and the negative effects of discrimination in the housing market and racism. However, change is occurring. Upwardly mobile members of ethnic minorities are gradually dispersing from central city locations to the suburbs and **exurbs** (Figure 6.7).

> **Exurbs** are settlements within a city region (villages, small towns), physically separated from the main city but functionally dependent on it for employment, retailing, leisure services, etc.

Central and eastern European immigrants have a more dispersed regional distribution. This pattern is a response to the:

● regional availability of employment
● wide range of skills and experience offered by them

One in five central and eastern European immigrants work in hospitality and catering, hence the large concentrations in the City of London and Westminster. One in seven works in labour-intensive agriculture and food processing, predominantly in eastern England and northeast Scotland. Other economic sectors such as administration and business, and health and medical care, are found at most levels of the urban hierarchy in the UK.

European immigrants are less segregated than South Asians and Afro-Caribbeans. As a general rule, the degree of segregation of an ethnic group is proportional to cultural distance between them and the host population. Even so, commonality of culture and language has attracted unusually large clusters of European immigrants to some towns, such as the Portuguese to Thetford (Norfolk) and Italians to Bedford.

Table 6.3 Leading employment sectors for central and eastern European immigrants

Employment sector	% employed
Admin, management, business	36.6
Hospitality and catering	19.0
Agriculture	10.8
Food processing	5.0
Health and medical	4.7
Retail	4.0

> **Examiner's tip**
>
> Scale provides a very handy structure to describe or explain a spatial distribution. Often the processes operating at a national or regional scale are different from those at a local or urban scale.

> **Typical mistake**
>
> It is a common mistake to forget that the distribution of ethnic minorities is dynamic. As Afro-Caribbean and South Asian families become better off and upwardly mobile, they move away from inner city areas, dispersing into the outer suburbs and commuter zone.

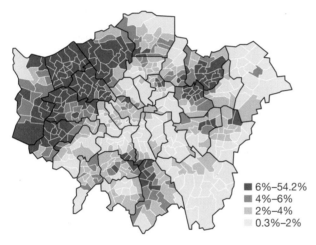

6%–54.2%
4%–6%
2%–4%
0.3%–2%

Figure 6.8 London's Indian population

Issues relating to multicultural societies

Revised

Conflicts often arise between the majority culture and ethnic minorities in multicultural societies. These conflicts are based on differences in values and traditions, and issues such as social and economic exclusion, separateness, inequality and fairness. Among the issues related to multicultural Britain are the following.

● Lack of integration by some ethnic groups (whether deliberate or the result of discrimination), especially those of Pakistani, Bangladeshi and African origin.

● Intolerance of cultural traditions/practices/values among ethnic minorities that conflict with the majority society (e.g. gender inequality, sharia law). Although honour killings and forced marriages are outlawed, their occurrence creates hostility and prejudice in the majority society. So, too, do cultural traditions such as large families and polygamy.

● The social and cultural domination of some inner city residential areas by ethnic minority populations (e.g. inner Bradford, Burnley and Oldham) leaves white British residents feeling isolated and alienated.

- Residential segregation (whether forced or voluntary) of ethnic minorities results in **ghettoes** which are often socially and economically self-contained. The outcome is ethnic minorities leading 'parallel lives', isolated and excluded from mainstream society. Segregated schools in ethnically diverse cities such as Bradford and Leicester reinforce separateness and lack of integration.

- The sheer volume of immigration to the UK in the 2000s was unprecedented. There was resentment by the host society that, at a time of high unemployment, jobs were taken by immigrant workers. Between 2001 and 2011, 1.77 million new jobs were created in the UK but 1.6 million were filled by immigrants, mainly from central and eastern Europe.

- High levels of unemployment among some ethnic minority groups, particularly among young men, and the economic exclusion this creates. The result is resentment, social tension and alienation. Ethnic riots in northern towns and cities in the 2000s were primarily economically motivated.

- Above-average levels of poverty and deprivation, particularly among Pakistani, Bangladeshi and African minorities and the impact on the overall cohesiveness of society. A vicious cycle of poverty and underachievement is driven by poor schools and inadequate command of English. For many second- and third-generation immigrants born in the UK, English is not their mother tongue. This is partly due to segregation and lack of integration, and partly to foreign-born parents who speak little English.

- Large numbers of bogus asylum seekers entered the UK in the 2000s who were not the victims of political persecution but were economic migrants. This imposes an economic burden on the rest of society.

> **Now test yourself**
>
> 7 What reasons underpin the Arab–Israeli conflict in the Middle East?
>
> 8 Outline the social, economic and environmental issues caused by the Arab–Israeli conflict.
>
> 9 What criteria define ethnic minority groups?
>
> 10 What is (a) multiculturalism, (b) a multicultural society?
>
> 11 What factors have contributed to the development of the UK's multicultural society?
>
> 12 Explain the uneven geographical distribution of ethnic minority groups in the UK.
>
> Answers on p. 134

Separatism

Separatism usually concerns ethnic minorities within a sovereign state that inhabit well defined geographical areas and demand greater control over their economic, social and political affairs.

The nature of separatism

Revised

The ultimate expression of separatism is the complete political withdrawal of a nation from a sovereign state. This is also known as **political secession**. An example is the secession of East Pakistan from West Pakistan in 1971 and the creation of the new sovereign state of Bangladesh.

However, full political independence is not always the aim of separatist movements. Lesser goals include greater autonomy, devolution and control of internal affairs. Separatism has long been an issue in French-speaking Quebec in Canada. But today, most Québécois seem to be satisfied that separatism has secured French as the official language of business and education in the province. The majority of Québécois show little interest in full independence.

Geographically, separatist movements are widespread. Even old democracies such as the UK and Sweden have separatist groups. Scotland's largest political party wants to create an independent Scotland, and the Sami people want their own autonomous region in northern Scandinavia. In newer democracies such as Spain, strong separatist movements exist in several regions, including Catalonia and the Basque Country. The Balkans in southeast Europe has been a region of diverse and conflicting ethnic groups for centuries. They were temporarily united with the formation of the Yugoslavia Federal Republic in 1918. But by the early 1990s, ethnic tension had resulted in violent conflict and the break-up of Yugoslavia. The outcome was a patchwork of new independent states including Bosnia and Herzegovina, Slovenia, Croatia, Serbia, Macedonia, and Montenegro (Figure 6.9).

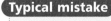
Typical mistake

Separatism exists in most sovereign states because states are rarely ethnically homogeneous and the boundaries of states and nations rarely coincide. However, most separatist movements are non-violent and few aspire to full independence.

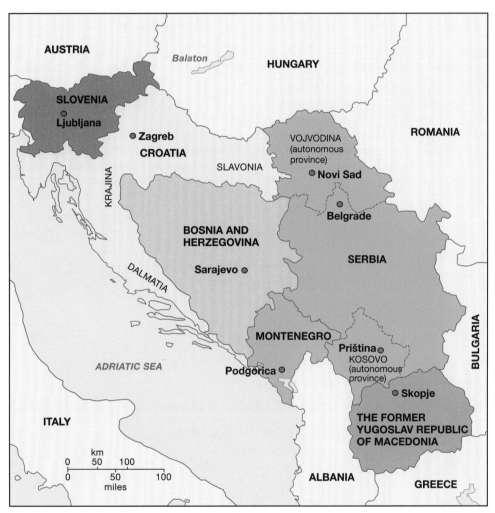

Figure 6.9 Yugoslavia and new sovereign states

Reasons for separatism

Revised

Separatist tendencies within a sovereign state are usually related to the political, economic and cultural oppression (real or perceived) of an ethnic minority by the majority population.

Examiner's tip

Separatism has many causes so it is important in examination answers not to overgeneralise but to refer to specific examples of separatism.

Table 6.4 Causes of separatism in Ulster, Kosovo and the Basque Country

Ulster	In 1923, Ireland (with the exception of Ulster or Northern Ireland) became an independent sovereign state. Ulster remained within the UK because of its majority Protestant (Unionist or pro-British) population. Although Ireland was partitioned along religious lines, a large nationalist (Catholic) minority remained in Ulster. During the next 40 years Protestants and Catholics lived separate lives — they were segregated residentially, each had their own schools, and social interaction between the two groups was minimal. The Catholic minority was systematically discriminated against. Unionists dominated the government and the police, few government contracts were awarded to Catholic businesses, and there was evidence of discrimination against Catholics in the allocation of social housing. These grievances led to the formation of a paramilitary nationalist organisation that aimed to unite Ulster with the Irish Republic. They were countered by similar unionist paramilitaries. Violent insurrection verging on civil war began in 1968 and continued for 30 years, claiming nearly 2,000 lives.
Kosovo	Kosovo was an autonomous province within the Yugoslav federation. After the break-up of Yugoslavia, Kosovo remained part of Serbia. Although only 10% of Kosovo's population were ethnic Serbians, Serbia had strong historical and emotional ties to the province, which was regarded as the cradle of Serbia's culture, religion and national identity. Separatism in Kosovo polarised around the distinct cultural identity of most Kosovans. Ninety per cent of Kosovans are ethnic Albanians. Their language is Albanian, not Serbian, and the majority are Sunni Muslims. As the pressure for full independence grew in the 1980s, the Serbian government reacted first by removing Kosovo's autonomous status and then by encouraging Serb settlers to move into Kosovo. In 1990 Kosovo's parliament was suspended and the Serbian government took control of the police, schools, medical facilities and the media. Albanian language newspapers and radio and television programmes were shut down, and Serbian became the province's official language again. Repression resulted in armed conflict. NATO forces fought alongside the Kosovan Liberation Army and eventually defeated the Serbs in 1999. Kosovo unilaterally declared itself independent in 2008, but the state has yet to be recognised by the UN.
Basque Country	The Basque Country extends from northern Spain across the border and into southwest France and is home to the Basque people. They speak an ancient language (Euskara) and have a distinctive rural culture and national identity. The Basque separatist movement arose out of the Spanish civil war (1936–1939) when Basque nationalists sided with the Spanish republic against a right-wing uprising led by General Franco, and declared autonomy. As a consequence of Franco's victory, the Basque leaders were either executed or imprisoned, the Euskara language banned and Basque culture suppressed. ETA, a Basque separatist and paramilitary organisation, was founded in 1959. Despite the death of Franco in 1975 and the move to democracy and autonomy for the Basque region, ETA continued its armed struggle for full independence until 2010.

Consequences of separatism

Revised

Separatism invariably leads to conflict. Sometimes conflict can be settled peacefully by negotiation and referenda, but often conflict erupts into violence, civil unrest and even war. The possible outcomes of separatism are the creation of a new sovereign state, greater autonomy with powers devolved from central government to a region, and the repression of minority groups that supported separatism.

Independence

Timor-Leste in southeast Asia, and South Sudan (see page 92) are twenty-first century examples of new sovereign states, created after prolonged and violent separatist struggles. East Timor, a former Portuguese colony, proclaimed independence in 1975, but within days was invaded and annexed by Indonesia. A long campaign of resistance against the occupying Indonesian army resulted in 100,000 deaths and three times as many refugees. Eventually, UN peacekeepers brought the war to an end and, in a UN-sponsored referendum, the East Timor people voted overwhelmingly for independence. East Timor was finally recognised internationally as an independent state in May 2002 and renamed Timor-Leste.

Regional autonomy

Separatist conflicts may be resolved by transferring most central government powers to a region. **Regional autonomy** stops short of full independence but is more wide-reaching than **devolution**. In Spain, the Basque Country and Catalonia have the greatest regional autonomy. The Basque parliament was set up in 1980 and has considerable fiscal and legislative autonomy. In addition to its responsibilities for healthcare, education and transport, the Basque Country has its own police force, collects taxes in the Basque region, has its own rate for corporation tax, and can vary the rate of income tax.

Devolution

Separatist movements may be appeased by devolving some power from central government. However, devolution stops short of regional autonomy. The Scottish parliament, set up in 1999, has powers devolved from Westminster over areas such as education, healthcare, environment and transport but has no tax-raising authority. It currently relies on an annual grant from Westminster. The Scottish National Party (SNP), the largest political party in the Scottish parliament, is committed to securing full independence for Scotland. Eventually the SNP plans to resolve the question of independence by referendum. However, in 2011 polls showed that only one in three Scots said they would vote for an independent Scotland.

Power sharing

After 30 years, 'The Troubles' in Northern Ireland ended with the Good Friday Agreement in 1998. The two main adversaries — the Unionists and the Nationalists — agreed to share power in a restored Northern Ireland Assembly. The Assembly has devolved powers over areas such as agriculture, regional development, health, education, justice and so on. Government ministers (11) are chosen by parties, their number being proportional to their popular vote. Thus in 2011 the Democratic Unionists had four ministers and Sinn Féin had three. The first and deputy first ministers are elected by the 108 members of the Assembly representing 18 constituencies.

Military defeat

In Sri Lanka, the Tamil Tigers, a separatist guerrilla group, conducted a prolonged war against the Sri Lankan government and the Sinhalese majority. The conflict lasted from 1983 to 2009 and claimed an estimated 80,000–100,000 lives. The Tigers' goal was to establish an independent Tamil state for the minority Tamil people. This aspiration was finally crushed by their military defeat in 2009. Two years after the end of the war the Sri Lankan government had made little attempt to address Tamil grievances. Although Tamil separatism will continue, armed conflict in future is unlikely. Ultimately this means a political solution and some devolution of power to the Tamil provinces in the north and east of the island. But at present stalemate prevails and Tamils may continue to suffer repression.

> **Examiner's tip**
>
> Separatism has a gradation of outcomes which could form the basis for an extended exam answer. The extremes are full independence and repression of an ethnic group. In between there is devolution, power sharing and regional autonomy.

The challenge of global poverty

The World Bank defines poverty as 'pronounced deprivation which comprises many dimensions'. Poverty includes, among other things, low incomes, poor health and education, poor access to clean water and sanitation, and insufficient capacity and opportunity to improve one's life. Extreme poverty is defined by the World Bank as living on less than US$1.25/day, moderate poverty on less than US$2/day. In 2010 more than half the population of sub-Saharan Africa (SSA) lived in extreme poverty. Globally the figure was more than 900 million.

The distribution of global poverty

The global distribution of poverty is heavily concentrated in the economically developing world, especially in SSA and South Asia (Table 6.5). Of the world's poorest countries (i.e. 35 low income countries where GDI per capita in 2010 was less than US$1,005), 29 were in SSA and South Asia (Figure 6.10).

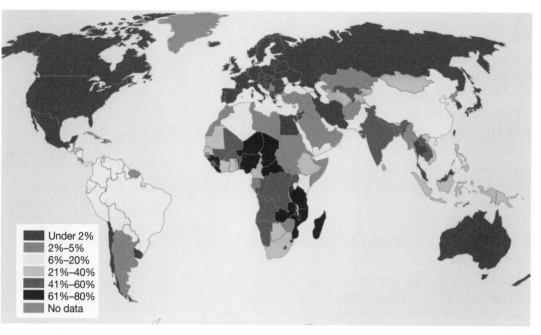

Under 2%
2%–5%
6%–20%
21%–40%
41%–60%
61%–80%
No data

Figure 6.10 Global poverty — percentage population living on less than $1/day, 2007–08
Source: UN Development Indices 2008

Table 6.5 Selected poverty indicators by region in the economically developing world

	Per cent living on US$1/day	Per cent children below minimum level of dietary consumption	Per cent literacy 15–24 year olds	Deaths before 5 years old per 1,000
North Africa	2.6	<5	86.6	26
SSA	50.9	26	72	129
Latin America & Caribbean	8.2	8	97	23
East Asia	15.9	10	97	19
South Asia	38.6	21	79.7	69
SE Asia	18.9	14	99.4	36
West Asia	5.8	7	93.2	32
Least developed countries	53.4	32	70.2	121

Causes of poverty

There is no single explanation for international differences in poverty. Poverty is due to a combination of physical, historical, political, social and economic factors.

Physical factors

- Several of the world's poorest countries are landlocked. Africa has 15 **landlocked countries**. Lack of direct access to the oceans restricts international trade and development, and contributes to poverty.

- Many LEDCs in the tropics and subtropics have arid and semi-arid climates. Drought is endemic (e.g. Niger, Somalia). Soils are often leached or saline and of low fertility.

Exam practice answers and quick quizzes at **www.therevisionbutton.co.uk/myrevisionnotes**

- Poverty is increased by the prevalence of tropical diseases such as malaria and yellow fever, which limit the productivity of labour.

Historical factors

- The legacy of **colonialism**. Colonies supplied food products and minerals for the benefit of the European imperial powers such as Britain and France. Economic development was encouraged only where it assisted resource exploitation (e.g. building ports or railways).
- Many African countries were the artificial creation of colonial powers. Their borders often ignored internal tribal, national and cultural differences. This situation contributed to political instability, separatism, civil war and poverty.

Political factors

Poor governance has hindered dozens of LEDCs in their drive to eradicate poverty. Corruption is often widespread — money that should be used for development and wealth creation is often channelled to the military and governing elite.

Social factors

- **Human capital** includes the education and skills of the workforce. It is a major factor driving development and improving living standards. In SSA more than a quarter of 15–24 year olds are illiterate and only a small minority completes secondary education.
- In many LEDCs tradition and religion discriminate against the education of women. **Gender inequality** neutralises a large part of the potential workforce, reducing productivity and increasing poverty.

Demographic factors

Almost all of the world's poorest countries have rapidly expanding populations with annual growth rates exceeding 2%. In extreme cases, population growth exceeds economic growth, exacerbating poverty.

Economic factors

- FDI is crucial to development, creating jobs that increase economic growth through spending, taxation and improvements in economic and social infrastructures. Large global inequalities exist in FDI. The EU, with 7.6% of the world's population, attracts 46% of all FDI. Africa, with 14% of the world's population, receives only 3.4% of FDI.
- There is also a close correlation between international trade, economic development and poverty. International trade is dominated by developed countries and contributes significantly to their prosperity. Africa and Asia, the poorest continents have 75% of the global population, but account for only 30% of world trade in merchandise and services.
- A prerequisite for successful economic development is a modern infrastructure of roads, railways, international airports, deep-water harbours and telecommunications networks. An efficient infrastructure is essential to attract FDI, tourism and many other economic activities.

> **Typical mistake**
> Inadequate human capital, gender inequality and rapid population growth act causally to hold back development and increase poverty. However, inadequate human capital is also the outcome as well as a cause of low levels of development.

Addressing poverty on a global scale `Revised`

International agencies such as the United Nations and World Bank are committed to reducing global poverty. The work of these multilateral agencies, governments in MEDCs, NGOs and progress in economic development has helped reduce the number of people living in extreme poverty. The proportion of the world population surviving on less than US$1.25/day should more than halve in the period 1990–2015, to less than 15%. The most spectacular progress has been made in rapidly industrialising countries such as China and India, but even in SSA, poverty levels should fall to just over one-third of the population by 2015.

Millennium Development Goals

Outlined in 2000, the UN's eight **Millennium Development Goals** (MDGs) set targets for global poverty, hunger, gender equality, education, environment and health to be met by 2015. Specifically the eight MDGs aim to:

1 eradicate extreme poverty and hunger

2 achieve universal primary education

3 promote gender equality and empower women

4 reduce child mortality

5 improve maternal health

6 combat HIV/AIDS, malaria and other diseases

7 ensure environmental sustainability

8 develop a global partnership for development

Although some of the MDGs will not be achieved by the deadline, significant progress has nonetheless been made. Overall there have been reductions in global poverty, improvements in primary school enrolment for girls as well as boys, reductions in child and maternal mortality and increasing HIV treatments (Table 6.6). However, while the proportion of poor people is declining, absolute numbers in SSA and South Asia are increasing, and lack of progress in reducing HIV is holding back improvements in both maternal and child mortality.

Table 6.6 Progress towards some MDGs in SSA and South Asia

MDG	Region	1990	Update
1 (% surviving on <US$1.25/day)	SSA	58	51 (2005)
	South Asia	49	39 (2005)
2 (% children completing primary education)	SSA	58	76 (2008)
	South Asia	79	90 (2008)
4 (under 5 mortality rate per 1,000 live births)	SSA	184	144 (2008)
	South Asia	121	74 (2008)

Development and security   Revised

In 2005 Kofi Annan, the UN Secretary General, made the memorable statement that there could be 'no development without security, and no security without development'. This sparked a debate among multilateral agencies, governments and NGOs about the effectiveness and targeting of international aid. Kofi Annan and the UN observed that violent conflict, poor governance and human development (e.g. reducing poverty, freedom from hunger, improving education and health) were interconnected (Figure 6.11), and that progress towards human development is impossible against a background of political instability and poor governance. Yet at the same time it is clear that poverty and deprivation spawn conditions of want and fear that allow terrorism, civil war and crime to flourish.

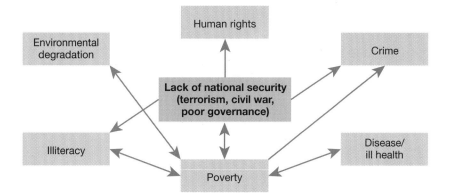

Figure 6.11 Peace, security and human development

Political insecurity causing lack of development

During the past decade multilateral agencies such as the UN, World Bank and IMF, and major donor governments have acknowledged that peace, political stability and good governance in LEDCs are prerequisites for development. In sub-Saharan Africa (SSA) civil wars and regional conflicts, such as those in Rwanda, DR Congo and Somalia, have cancelled out years of development and progress. Resources that should have been devoted to development have instead been squandered on war and armed conflict. In Somalia, 20 years of civil war have created lawlessness, anarchy and a state without a government. Development has gone backwards and widespread famine in 2011, caused as much by civil war as drought, led to tens of thousands of Somalis fleeing to refugee camps in neighbouring Kenya.

Given the chaotic political conditions in Somalia it is impossible to direct long-term development aid there. Although Somalia is an extreme case, similar conflicts that place the effectiveness of development aid in doubt, are widespread in SSA. They include on-going wars in Ivory Coast, Sudan, DR Congo, Chad and Nigeria.

But countries wracked by war, insurrection and poor governance are also unlikely to be attractive to TNCs and inward investment. Many experts would contend that social development is best achieved not through aid programmes but by investment in industry and physical infrastructure, which generate profit and enable governments in LEDCs to build schools and hospitals. This can only happen where there is peace and political stability.

Lack of development causing political insecurity

Figure 6.11 shows how inadequate and unequal human development is also a driver of political instability in LEDCs. Poverty, hunger, poor health and education, oppression and infringements of human rights create resentment and conflict. Where these problems coincide with existing separatist tendencies, the outcome may be armed conflict, civil war and political instability. Thus, it is argued that to achieve security, peace and good governance, human development issues need to be tackled first.

Ultimately development and security cannot be viewed in isolation. They are part of a complex, interconnected system. Poor levels of development create conditions that lead to insecurity. But equally, with an absence of security, social and economic development cannot take place. Should development programmes give priority to national security (the UN has, after all, always performed an important peace-keeping role) or to economic and human development? And what is the best way to break this vicious circle of political insecurity and lack of development?

> **Examiner's tip**
>
> Discussion of this issue will clearly invoke judgements. Links between security and development are difficult to establish objectively, so arguments must be supported with actual examples. There should also be awareness that international aid designated for security (i.e. to create a peaceful environment for development) could be abused by recipient governments.

Conclusion

While multilateral organisations and many world leaders believe that development can only be effective with security and peace and *vice versa*, there are voices of dissent. Some NGOs fear that this belief will encourage governments in LEDCs to (a) divert international development aid to military spending to defeat opposition and separatist groups, and (b) crack down on political opposition and dissent. This in turn could escalate conflict, violate human rights and set back the whole development process.

Now test yourself

13 Describe the distribution of poverty at the global scale.

14 Draw a diagram to show how the global distribution of poverty is caused by a range of different factors.

15 What are the Millennium Development Goals (MDGs)?

16 Why are many MDG targets unlikely to be achieved by 2015?

17 (a) How does a lack of national security restrict development?
   (b) How does a lack of development contribute to national insecurity?

Answers on p. 135

Check your understanding

1 Assess the view that the Arab–Israel conflict in the Middle East is based on conflicting claims on territory and resources.

2 Why do the boundaries of sovereign states and nations rarely coincide?

3 To what extent are separatist movements influenced by geography?

4 What factors give rise to environmental issues at a local scale in MEDCs and how are these issues resolved?

5 What are the economic and social advantages and disadvantages of large-scale immigration to a country?

6 What factors contribute to the geographical segregation of ethnic minority groups?

Answers on p. 135

Exam practice

Section B

1 (a) Study Figure 6.10 which shows the global distribution of poverty. Comment on the global distribution of poverty in Figure 6.10.   [7]

   (b) Using examples, suggest reasons for the geographical patterns of poverty at a global scale.   [8]

   (c) How successful have international agencies been in tackling the issue of global poverty?   [10]

2 (a) Study Figure 6.5 which shows international migration trends in the UK between 1991 and 2009. Comment on the main features of international migration in the UK in Figure 6.5.   [7]

   (b) Describe and explain the distribution of one major ethnic minority group in the UK.   [8]

   (c) Discuss the influence of geography on the economic and social issues associated with the UK's multicultural society.   [10]

Section C

3 Analyse the reasons for the Arab–Israel conflict and discuss its geographical consequences.   [40]

4 Discuss the view that there is 'no development without security, and no security without development.'   [40]

Answers and quick quiz 6 online

Online

Examiner's summary

✔ Discussion of conflict should consider geographical scale as a useful structure for analysis. General statements for discussion may apply at some scales but not others.

✔ In many international conflicts ideology may hide the true motives of the protagonists, which often include controlling resources and extending a country's power and influence.

✔ Local conflicts are usually resolved by enquiry, discussion and debate. But, however objective the evidence, outcomes are determined by decisions that are subjective and based on an individual's or society's value systems.

✔ In local environmental conflicts, the protected status of an area does not ultimately guarantee immunity from development.

✔ Discursive exam answers require students to make judgements on issues based on the strength of evidence, filtered by their own values.

✔ Transnational environmental issues are often centred on resources (e.g. water, unpolluted air) shared by neighbouring states.

✔ Multiculturalism and immigration are often associated with sensitive issues that must be discussed fairly and objectively. Judgements can still be made, but they must be free from prejudice and any ethnic bias.

✔ The geographical distribution of minority ethnic groups at regional and local scales in MEDCs is dynamic. Counterurbanisation affects ethnic minorities as well as the majority population. The idea that ethnic minority groups are confined to inner city locations is a stereotype and is out of date.

✔ Extended answers must avoid excessive generalisation and students must support claims and illustrate discussion with case studies and/or examples.

✔ The distinction between cause and effect can be ambiguous. The poor quality of human capital in a country may be both the cause and the result of lack of development. Similarly lack of security and poor governance can be a major obstacle to human development, but dissatisfaction with levels of development and governance can spark violent conflict.

7 Geographical fieldwork investigation

Introduction

Fieldwork investigation is an important route to the acquisition of geographical data. The successful execution of a fieldwork investigation demands a range of skills that cover areas such as planning, data collection, data presentation, data analysis and evaluation (Figure 7.1).

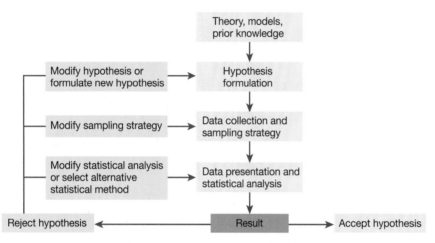

Figure 7.1 Geographical investigation

As part of the AQA specification you may opt to complete a personal fieldwork investigation based on **primary data**. The topic chosen must be derived from the content of the A-level specification. This connection with other parts of the specification, together with the use of geographical skills, helps to satisfy the requirement for **synopticity**. The investigative study is assessed directly in Unit 4A through a written examination, which also includes questions of a more general nature on fieldwork skills (see Geographical Skills section in the AS Revision Guide). In this section we focus on revision themes you need to address in relation to your personal investigation.

> **Synopticity** is defined as understanding the connections between different aspects of the geography in the specification and demonstrating an ability to 'think like a geographer'.

Geographical fieldwork skills

Purpose of the investigation
Revised

The investigation should have a *single* purpose or aim. This might be to:

- test a **hypothesis**
- answer a question
- investigate an issue or problem

Justification of the purpose or aim of the study will rely on the relevant geographical and conceptual (theoretical) background. Thus the context

> A **hypothesis** is a statement about a presumed relationship or difference between variables whose validity can be tested through scientific (or geographical) investigation.

for an investigation of river channel width and flow velocity would include channel efficiency (i.e. width/depth ratio, hydraulic radius) and the increasing loss of energy to friction in wider, shallower channels. Investigations in physical geography are most likely to be underpinned by a body of theory. When an economic, social or environmental issue is investigated, relevant context would focus on the background to the conflict (e.g. origin/history, protagonists, arguments).

Examiner's tip

You must have a clear understanding of the purpose of your investigation and be able to present it in the context of relevant theory or background.

Planning and data collection

Revised

The planning stage requires consideration of a number of important questions.

- Where is the fieldwork to be conducted, when, and why was this location chosen?
- What are the potential safety issues and risks involved with fieldwork in this location?
- What is the scale of the investigation and how long will it take to collect the **primary data**?
- How much data will be collected and in what form?
- How will the data be collected and what precautions will be taken to ensure their accuracy?

Location for fieldwork

The location chosen for fieldwork should provide the amount and type of data required. For instance, a small town might be chosen for an investigation into shopping behaviour because it has a good range of shops, has little competition from other centres in the region, and a market that attracts a large number of shoppers on market day. A river might be chosen because it is easy to access (on public land, student had the landowner's permission), has a coarse (rather than sandy) bedload, has a channel free from dense vegetation, is little affected by human activity, and is a safe environment in which to work.

Primary data are new data that have not previously been collected and processed (e.g. fieldwork data). Secondary data usually comprise published documents such as textbooks, articles, internet files, etc.

Examiner's tip

You must be able to provide coherent reasons why you chose a particular location for your study.

Safety issues and risk assessment

Geographical fieldwork always includes an element of risk. Precautions must be taken to minimise risk by (a) identifying potential hazards, (b) assessing the risk posed by these hazards, (c) devising a plan or strategy for dealing with the hazards. This process is formalised by completing a risk assessment audit, which is submitted to a teacher or supervisor for approval.

Depending on the location and the nature of the fieldwork, common safety precautions include:

- working in groups of at least three people
- carrying a mobile phone with essential contact numbers (e.g. supervisor's number)
- wearing suitable outdoor clothing and footwear for protection against wet and cold weather and rough terrain
- carrying first-aid equipment, and where appropriate a torch, whistle, survival bag, emergency rations and a detailed map (1:25,000) of the study area

Examiner's tip

Be prepared for exam questions that require you to identify potential hazards in your chosen fieldwork location and the action you took to minimise them.

● leaving precise details of your itinerary with your teacher or supervisor, including times of departure and return

Data collection

Quantity of data. Most investigative studies are based on a small part of the **statistical population** known as a **sample**. Sample size will depend on (a) the time and resources available for data collection, (b) the level of precision and confidence required in the final outcome. Thus the amount of data collected will be the minimum needed to provide accurate and reliable results.

Form of data. The form of the data (e.g. quantitative or qualitative) will determine how they can be used. If you decide to analyse data statistically, the data must be in quantitative form. If the approach to investigation is more qualitative, information comprising descriptions of people's feelings and opinions may be appropriate.

Representative and reliable data. Data collection needs to be objective to ensure its reliability and representativeness. Choice of sampling strategy is crucial and must be justified. Equally important is the method of data collection, whether by observation, measurement or interview (questionnaire).

Random sampling is practicable where a listing of the population exists. For example, households listed in the electoral register could be selected for interview using random numbers. **Systematic sampling** selects individuals at regular intervals (e.g. every *n*th household or individual, or every *n*th metre). It is versatile, quick and well suited to most surveys in both physical and human geography. **Stratified sampling** is used where a population comprises a number of subgroups (e.g. zones of sediment on a point-bar, households classified by income). Once the subgroups are identified, individuals are selected either randomly or (more often) systematically.

Location is an important dimension of the population under review in most geographical investigations. In these circumstances **spatial sampling** is used. Samples are selected either randomly or systematically at **points**, in areas (e.g. **quadrats**) or along **lines of transect**.

Data presentation

In order to generalise data and highlight patterns and trends, data are presented as tables, charts and maps.

Tables   Revised ☐

Tables are the simplest method of presenting data. Data are generalised into groups and classes and presented in a matrix of columns and rows. If patterns and trends are clear there may be no need for more elaborate presentation.

Charts

Charts should be chosen on the basis of their clarity, accuracy, visual effectiveness and the characteristics of the data. Possible justifications for using charts are as follows.

- **Histograms** — where data are in the form of frequencies.
- **Bar charts** — where data relate to discrete places or units of time, e.g. mean discharge on a river recorded at several gauging stations downstream. **Stacked bar charts** are used to display subsets of data, e.g. age structure in a country or region divided into broad age groups (0–14, 15–39, 40–65, over 65 years).
- **Pie charts** — an alternative to stacked bar charts. Normally they are used to represent proportions or percentages.
- **Line charts** — used where data are continuous in time and space, e.g. mean daily or hourly flow on a river, average urban population densities with distance from the CBD.
- **Dispersion diagrams** — single axis charts used to compare the distribution of variables from two or more data sets.
- **Scatter charts** — used to plot two variables, x and y, where x is the **independent variable** causing change in y, the **dependent variable**.
- **Triangular charts —** used to plot three variables on a percentage scale, whose values sum to 100% (e.g. proportion of children, adults and old people in a population). They provide a visual comparison of difference between places, as well as changes through time.

Chart scales can be arithmetic or logarithmic. **Logarithmic scales** are chosen where (a) a range of values is too great (or too narrow) to plot clearly on an arithmetic scale, (b) where the relationship between two variables is geometric (curvilinear). Plotting one or both variables on a logarithmic scale converts a curvilinear trend to a straight line.

> **Typical mistake**
>
> While it is relatively easy to describe a chart, students often struggle to give reasons why a particular chart was chosen to present data.

Maps

Statistical maps relate data to specific locations and show spatial trends and patterns more effectively than charts.

- **Dot maps** show detailed spatial patterns *within* areal units, such as census areas. Unlike choropleth maps, distributions are not interrupted by artificial administrative boundaries. Dot maps also provide an excellent visual impression of distributions and densities. They are, however, time-consuming to draw and the placement of dots is often highly subjective.
- **Choropleth maps** are often used with data that are aggregated for areal units, such as postcode areas, census areas and local authority areas. They are most useful in plotting data standardised by area — for instance, population densities and cropland as a proportion of total cultivated land. Their main drawbacks include excessive generalisation of data where areal units are large, distorted spatial patterns when areal units vary enormously in size, and sudden and unrealistic changes in values across areal boundaries.

- **Proportional symbol maps** use circles, squares, triangles and other shapes to represent values proportional to the area of the symbols. They show absolute, not standardised values and therefore may be used to plot variables such as population counts, employment and retail floorspace. Circles can be subdivided like pie charts to map additional information. The main weakness of proportional symbol maps is estimating symbol areas — doubling the sides of a square or the radius of a circle increases the area of the symbol four times, not twice.

- **Isoline maps** use isolines or isopleths to join places of equal value. They are often used in physical geography to show distributions that are continuous in space, such as rainfall (isohyets), pressure (isobars) and temperature (isotherms). There are also instances when they are used effectively in human geography, for example to plot land values or pedestrian densities in a CBD.

- **Flow maps** are the only maps that show the movement of people, goods and information between places. Flow paths can be either routed or non-routed, the latter giving rather more detail about traffic.

Data analysis

Data analysis will usually involve the use of at least some simple descriptive statistics. Where the data permit, more advanced analysis using inferential statistical techniques are used.

Descriptive statistics

Data sets are often summarised by a single representative value. There are three possibilities: **arithmetic mean**, **median** and **principal mode**. Possible reasons for using one or more of these **measures of central tendency** include:

- data sets with few extreme values that cluster around a central value and which follow a normal frequency distribution are accurately summarised by the arithmetic mean

- the median (or middle value) is a good descriptor when data sets contain extreme values and have a skewed (rather than normal) distribution

- the principal mode is the class in a histogram or frequency table that has the most values. It can be a useful tool where only broad descriptions of data sets are needed

Measures of dispersion summarise the scatter of values in a data set around a central value, such as the mean or the median. The simplest measure of dispersion is the **range**; but based on two values it conveys only a limited amount of information. The **inter-quartile** range, employed with the median, uses half the values in a data set and is therefore more representative. However, the **standard deviation**, calculated from the full data set and used with the mean, provides the most detailed and accurate description. When comparing the dispersion of values in different data sets, the standard deviation is often expressed as a percentage of the mean. This statistic is known as the **coefficient of variation**.

Inferential statistics

Revised

Spearman Rank correlation coefficient, the **Mann-Whitney U test**, and the **Chi-Squared test** are inferential statistical tests. They determine whether the relationship or difference between variables is **statistically significant** — in other words, when there is 95% confidence that the relationship or difference is not due to chance.

- Spearman Rank correlation is used to test hypotheses of the form: 'x *is related to* y'. It can be applied to small data sets where data have an ordinal or rank scale. The outcome is a coefficient of correlation which ranges from −1 to +1, where −1 is a perfect inverse relationship, +1 is a perfect positive relationship, and 0 is no relationship.

- The Mann-Whitney U test is applicable to hypotheses of difference, i.e. is A *different from* B? Like Spearman Rank correlation it uses ordinal data that relate to two variables. It is also effective on small data sets (e.g. as few as ten pairs of values).

- Chi-Squared is used to determine whether an observed frequency distribution differs significantly from frequencies that might be expected if the distribution were random. It is applied to (a) hypotheses of difference, (b) data that are in the form of frequencies (it cannot be used for percentages of proportions), (c) data sets that have few categories which contain less than five expected values.

> **Typical mistake**
>
> Poor understanding of the concept of statistical significance and the application of inferential statistics in analysis of sample data are common errors. It is important to know why a particular type of statistical test has been used.

Completing the investigation

Conclusions and evaluations

Revised

Investigations are completed by (a) providing a summary of the main findings, and (b) evaluating the outcomes.

Students should be prepared to give an outline of their main findings in the examination. They should also recognise that geographical investigation is often problematic and that outcomes may be ambiguous and subject to a number of caveats. Evaluation of the effectiveness of the enquiry might consider some or all of the following questions.

- Was the hypothesis or question that formed the focus of the investigation appropriate, or with hindsight should it be modified?
- Were sufficient data collected and did the sampling method prove robust and provide accurate and truly representative data?
- Was the scale of the study appropriate?
- What are the limitations of the enquiry and how could it be improved and/or extended?

In answer to these questions, possible alternatives should be discussed. Finally, some discussion is needed on the extent to which the initial aim(s) of the investigation have been achieved. In this context investigations that have been properly planned and executed but which ultimately reject a hypothesis (i.e. a negative finding) are, of course, just as valid as those that verify a hypothesis.

> **Now test yourself**
>
> 6 Describe and justify any (a) charts, (b) mapping methods you used in your personal study to represent primary data.
>
> 7 Explain how you analysed the primary data in your personal study. Give reasons for the techniques you used.
>
> 8 Make a list of the outcomes of your investigation.
>
> 9 Assess the extent to which you feel that your investigation achieved its aim(s).
>
> **Answers on p. 136**

Check your understanding

1 Describe the main stages of investigation of a geographical hypothesis or problem.
2 Describe the main methods of statistical sampling and assess their usefulness.
3 Critically assess the value of (a) dot maps, and (b) choropleth maps to show the distribution of population in an area the size of an urban or rural district.
4 Describe three measures of central tendency and assess their usefulness.
5 (a) Explain what is meant by statistical significance, (b) describe one inferential statistical technique and explain the circumstances when it would be used.

Answers on p. 136

Exam practice

Section A

State the aim of your investigation.

1 Describe the factors that influenced your choice of location of your fieldwork investigation. [12]
2 (a) Describe the methods you used to collect your fieldwork data. [6]
   (b) Explain the steps you could take to improve the quality of the fieldwork data you collected for your investigation. [12]
3 Discuss the extent to which (a) your investigation achieved its aim, and (b) your findings are consistent with relevant theories and concepts in your field of study. [10]

Section B

4 Data from a fieldwork investigation into the influence of slope angle and drainage on the distribution of bracken in a moorland area in northern England are shown in Table 1.

Table 1 Slope angle and bracken distribution — Muggleswick Common, Co Durham

Slope categories (°)	Per cent area occupied by each slope category	Number of metre quadrats dominated by bracken
0–6.9	44	4
7–13.9	38	18
>13.9	18	11

The data were used to test the following hypothesis — the dominance of bracken increases as slopes steepen.

Analysis of the data using the Chi-Squared test produced a Chi-Squared value of 14.31. Critical values for Chi-Squared with 2 degrees of freedom are given in Table 2.

Table 2

df	0.10	0.05	0.01	0.005	0.001
2	4.61	5.99	9.21	10.6	13.82

(a) Using the Chi-Squared value calculated, what conclusion can be made about the validity of the hypothesis? [4]
(b) Suggest and justify an alternative way of presenting the data in Table 1. [4]
(c) Explain how statistical analysis of fieldwork data can increase geographical understanding. [12]

Answers and quick quiz 7 online

Online

Examiner's summary

✔ Prior to sitting the Unit 4A examination, students must have a thorough knowledge and understanding of their personal investigative study, including its aim, theory and context, methodology and outcomes.

✔ Sound reasons are needed to justify choices of sampling methods, data presentation and data analysis techniques, as well as an understanding of possible alternatives.

✔ Consideration must be given to the extent to which the personal investigation achieved its aim.

✔ There should be a critical appreciation of the limitations of the personal investigation and ideas on how, with hindsight, it could be improved.

✔ Where statistical methods have been used, there must be a clear understanding of the concept of statistical significance.

8 Geographical issue evaluation

Introduction

Content and assessment — Revised ☐

Unit 4B is an alternative to Unit 4A. It assesses the same range of geographical skills but through an issue evaluation exercise rather than a personal investigation and a written examination of geographical methodology. Most of the knowledge covered in the previous chapter is directly relevant to Unit 4B and should be studied closely.

Advance Information Booklet (AIB)

Unit 4B is based on an AIB released on 1 April and 1 November for each series. The AIB provides resources that relate to a specific geographical issue. They include text, documents, tables, charts and maps. The issue may have social, economic, political and environmental dimensions and is connected to topics studied elsewhere in the A-level specification.

Synopticity

Inter-connections between issue evaluation in Unit 4B and topics studied at AS and A2 fulfil the synoptic requirement for this part of the specification. However, it should be noted that only two of the four topics studied in Unit 1 (AS) are core topics, and that all topics studied in Unit 3 (A2) are optional. The two AS core topics are Rivers, Floods and Management, and Population. Synopticity is also provided by Geographical Skills (Unit 2) which is a compulsory part of the specification.

Geographical issue

The geographical issue presented in the AIB will involve conflict between various interested parties such as international agencies (e.g. UN), national and local governments, government agencies (e.g. Environment Agency), business interests, non-governmental organisations (e.g. charities) and individual people. The main task is to evaluate this issue. This means scrutinising all of the evidence; assessing the strength of the arguments, strategies and proposals; and making judgements. Although evaluation must be logical, consistent and based on evidence, it will be influenced by the personal values, priorities and experience of the student.

Preparing for the examination

Once the AIB has been released, preparation for the written examination begins. This work requires systematic revision and research of several key themes.

Interpret the data and resources provided by the AIB

● Describe and summarise in note form the main trends and patterns shown in tables, charts and maps. Assess their contribution to, and their significance for, the issue under consideration.

● Identify the arguments in text and documents that are relevant to the issue, and the positions and attitudes adopted by interested parties.

Use techniques to present and analyse data from the AIB

● Select appropriate graphical, cartographical and statistical methods (see Chapter 7) to represent data in the AIB.

● Justify your selections, using criteria such as accuracy, visual effectiveness, subjectivity, relevance to the data and ease of construction.

● Revise the techniques of construction used in the charts and maps chosen (Unit 2).

Suggest additional information

Sources and types of additional information will depend on the nature of the geographical issue in question. Possible methods of collecting additional primary information include questionnaire surveys, street interviews, land-use surveys, environmental quality surveys, measurements of channels, sediments, gradients etc. Spatial sampling strategies might use quadrats, points or transects. Secondary information at a local or regional scale might be obtained from websites such as the National Statistical Office (neighbourhood statistics), the Environment Agency and local authority sources. Google Earth provides an excellent visual impression of the locational characteristics of the area affected by an issue. At national and international scales, websites such as the International Data Base (US census bureau), UN population data base, FAOSTAT and so on can provide useful background information.

Recognise shortcomings of the data

● Data may be out of date (e.g. may refer to census data more than 4 or 5 years old).

● Data may not be appropriate for a local study (e.g. refer to wards and parishes rather than the smallest census areas, i.e. Lower Super Output Areas).

● Data may initially have been collected for other purposes (rents used as surrogate data for land values).

● Data may lack accuracy, e.g. derived from unrepresentative samples or small samples.

● Data may be in nominal and/or ordinal rather than ratio form and therefore inappropriate for some data presentation methods and statistical analysis. Data may be standardised (e.g. percentages) rather than absolute values.

● Investigate other sources of data through internet research in order to satisfy some of these shortcomings.

Relate the data to your AS and A2 studies

This instruction is a reminder of Unit 4B's synoptic function. The Unit cannot be considered in isolation — it also requires knowledge and understanding of topics relevant to the AIB theme studied at AS and A2. However, it is worth noting that only Rivers, Floods and Management, and Population in Unit 1 at AS are core topics (i.e. studied by all students). The Geographical Skills Unit at AS is also a core area and revision of this topic is vital. All six topics in Unit 3 are optional. Themes which fall outside the student's own study area should be researched in the period between the release of the AIB and the examination.

Geographical issues

Revised

Recognising and defining issues

Issues are situations and actions that:

- polarise opinion and are controversial
- cause conflict between interested parties, e.g. local people, businesses, conservationists, etc.
- imply social, economic and environmental change

Situations around which issues develop may be recognised by possible social and economic inequalities, exclusion of some socio-economic and/or ethnic groups, environmental problems that may affect communities unequally, environmental degradation that conflicts with conservation.

Considering evidence from different points of view

Most issues involve several groups or 'actors', each with their own agendas and attitudes to change. Issues must be seen from these different perspectives to appreciate their potential impact and evaluate the legitimacy and strength of arguments.

Establishing criteria for issue evaluation

Geographical issues can be evaluated against a number of criteria. However, the weighting and importance given to each criterion will be subjective and reflect a student's own value system. Possible evaluative criteria include:

- the cost of change, measured in economic, social and environmental terms
- the scale of social, economic and environmental change
- the benefits of social, economic and environmental change
- the impact of change measured against formal policies and objectives
- long-term and short-term costs and benefits
- the economic and environmental sustainability of change
- national versus local interests
- the extent to which change is likely to produce social and economic inequalities

Managing conflict

Evaluate a range of options

Evaluating management responses to an issue requires a thorough understanding of the issue and its implications. Researching the issue with reference to other areas of the specification and to new resources from textbooks, articles and the internet, is essential. Evaluation of management options might consider their location, feasibility, cost, social/economic/political/environmental impact, sustainability and inequity.

Resolving and reducing conflict

Conflicts could be resolved by:

- negotiation and compromise, which ensures that costs and benefits are distributed equally
- formal planning procedures such as public enquiries
- unilateral decisions taken by non-democratic governments
- compensating injured parties

Examiner's summary

- ✔ In the period between the release of the AIB and the examination, detailed background research on the issue should be undertaken.

- ✔ Students should have an awareness of the synoptic requirement of Unit 4B and revise all aspects of the AS and A2 specifications (knowledge and skills) that relate to the issue in the AIB.

- ✔ Data provided in the AIB should be viewed critically (accuracy, amount, representativeness, etc.)

- ✔ The choice of techniques to present and analyse data should be justified.

- ✔ Issues should be evaluated against a checklist of relevant criteria.

Answers

Chapter 1

Now test yourself

1 Oceanic crust comprises basalt, is relatively thin (5–10 km) and is less than 200 million years old. Continental crust is made of granite, is up to 70 km thick and may be several billion years old. Continental crust is lighter (i.e. less dense) than oceanic crust.

2 Check your diagram against Figure 1.1 on p. 7.

3
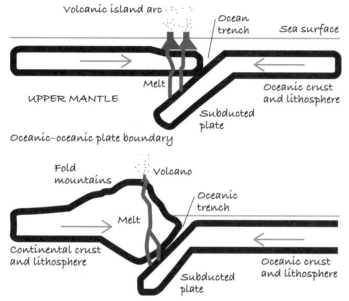

4 Any three of the following:
   • the matching shapes of the continents on either side of the North and South Atlantic Ocean, which suggest that they might once have been joined
   • ancient glacial deposits in the southern continents, formed during a past glacial period but now widely dispersed
   • the same species of fossil plants and animals found in modern Africa and South America
   • similar rock types and structures on opposite sides of the Atlantic in Brazil and west Africa, and northwest Europe and northeast North America

5 The polarisation of iron particles in the oceanic crust. Periodic reversal of the poles causes iron particles in molten rocks to be aligned alternately towards the north and south poles. These patterns of magnetic striping are mirrored on either side of the mid-Atlantic ridge, proving that new crust formed here and spread laterally away from the ridge, i.e. sea-floor spreading.

6 (1) Effusive eruptions — quiet, non-violent eruptions of basaltic, non-viscous magma which forms lava flows, (2) explosive eruptions — violent eruptions associated with viscous (andesitic) magma which traps gases such as steam, and produces pyroclastic flows.

7 Violent eruptions occur when thick, viscous magma does not allow the escape of gases such as steam. These build to enormous pressure and often result in volcanic explosions.

8 Check your diagram against Figure 1.8 on p. 12.

9 (1) Lava flow — hot, liquid rock extruded from a volcano or fissure. (2) Pyroclastic flow — fast-moving, fluid-like flow of hot ash, boulders and gas caused by a violent eruption. (3) Lahar — mudflow or debris flow originating on a volcano. (4) Ash fall — tiny rock fragments and pumice, emitted into the atmosphere in dense clouds, which fall back to the surface under gravity, blanketing the ground in debris.

10 Early warnings to communities under threat of eruption by: (1) monitoring earthquake activity within volcanoes, (2) measuring ground inflation, (3) monitoring the gases emitted by the volcano, (4) evacuation of populations at risk. In some circumstances lava flows can be diverted away from settlements.

11 Inter-plate earthquakes occur at the boundaries of tectonic plates and are often associated with subduction and rifting. Intra-plate quakes occur at locations remote from plate boundaries and result from earth movements along fault lines.

12

Preparedness	Earthquake-proof/fire-proof buildings, disaster planning
Magnitude	Energy released by the quake — the most powerful quakes are magnitude 8–9
Ground material	Liquefaction, where ground comprises water-saturated sediments
Location	Low-lying coastal areas exposed to tsunamis generated by submarine earthquakes
Poverty	Poor construction (houses, infrastructure), lack of resources to survive the aftermath of an earthquake
Time	Earthquakes that occur at night have a greater impact than those in the daytime

13 Primary hazards — ground shaking, liquefaction; secondary hazards — tsunamis, landslides, debris flows, fires, disease, etc.

14 (1) Construction of earthquake-proof buildings and infrastructure (building codes, deep foundations, reinforced structures, fire-resistant structures, etc.). However, this is costly and beyond the resources of most LEDCs. (2) Disaster planning to provide emergency relief, a strategy for long-term recovery and reconstruction. These responses assume low levels of priority in poor countries that rely heavily on international aid.

15 Tsunamis are mainly triggered by undersea earthquakes, often along subduction zones such as those along the Pacific coast of Japan, and Sumatra's Indian Ocean coast. Tsunamis are particularly devastating where (a) undersea earthquakes occur near to land, (b) coastal areas exposed to tsunamis are low-lying, (c) exposed coastal areas have high density populations.

Check your understanding

1 The Earth's crust and lithosphere are in continuous motion, driven by the circulation of convection currents in the plastic upper mantle or asthenosphere. New crust and lithosphere form at mid-ocean ridges (constructive plate margins). Over millions of years it flows laterally like a conveyor belt until, finally, it subducts along destructive plate margins. The result is continuous recycling of the oceanic crust and lithosphere. In this system, rates of formation and destruction are perfectly balanced. Recycling extends over a period of approximately 200 million years.

2 The circumstantial evidence for continental drift includes: matching geological structures on opposite sides of the Atlantic Ocean (e.g. Scandinavia/Scotland and eastern Canada/USA); the matching shapes of the continents on either side of the Atlantic Ocean; the same fossil plant and land animal species found in Africa and South America; and similar glacial deposits in Africa, South America and Antarctica. The most powerful evidence is the mechanism for continental drift, i.e. sea-floor spreading.

3 Volcanic activity is mainly concentrated on or close to constructive and destructive plate boundaries (exceptions are hot spots such as Hawaii, Yellowstone and the Canary Islands). At constructive plate boundaries plumes of magma rise to the surface. As pressure falls and the magma rises, volcanic activity occurs on the ocean floor along the mid-ocean ridges. In places volcanic activity is so intense that volcanoes rise above sea level to form islands (e.g. Iceland, Tristan da Cuhna). At subduction zones volcanic activity is caused by the destruction of the crust/lithosphere. As the crust/lithosphere descends to the mantle it melts at depth. This melt, which is lighter than the surrounding rock, slowly rises towards the surface where it forms chains of volcanoes. In the ocean these volcanoes often form island arcs (e.g. Lesser Antilles). On continents they may form clusters of volcanoes often embedded within fold mountain ranges (e.g. the Cascades).

4 Major earthquakes often occur at inter-plate boundaries (e.g. the 2011 earthquake off the east coast of Japan which triggered a massive tsunami). Inter-plate earthquakes are mainly due to enormous pressures and tensions that develop at plate boundaries. At destructive and conservative boundaries plate movement is often temporarily halted by friction. As pressures build, rocks eventually snap causing sudden (and violent) movement, which releases huge amounts of energy as shock waves. At constructive plate boundaries where the crust and lithosphere are being stretched, extreme tension causes rifting (i.e. vertical fault lines) which produces crustal movement and earthquakes.

5

	Mitigating action	Effectiveness
Volcanic eruptions	Monitoring volcanoes — seismic activity, inflation, gaseous emissions, etc.	Highly effective in forecasting the likelihood of eruption but unable to predict the exact timing of an eruption.
	Evacuation of population from areas most at risk.	Normally effective, though depends on the explosiveness and direction of the blast.
	Diversion of lava flows.	Has occasionally been successful, e.g. Mount Etna, Heimaey, Hawaii.
Earthquakes	Implement building codes to make buildings earthquake proof (e.g. massive steel frames, shock absorbers, fireproofing, etc.)	Highly successful in earthquake-prone regions in rich countries, e.g. Japan, USA. Limited effectiveness in LEDCs where poverty often means that building codes are not enforced and the resources to make buildings earthquake-proof are not available.
	Planning for disaster by governments — education of population, responses by emergency services, availability of food, water, temporary shelter, etc.	Highly successful but requires sophisticated organisation often only available in MEDCs.
	Responses to earthquake disasters by foreign governments and multilateral agencies providing emergency relief aid.	Most LEDCs depend heavily on the international community to provide emergency relief aid. Examples include the recent earthquakes in Kashmir and Haiti.
Tsunamis	Tsunami warning systems and evacuation planning.	Early warnings following an earthquake may give time for evacuation. Their effectiveness depends on the proximity of any coastal area to the quake epicentre.
	Construction of coastal defences such as seawalls.	Costly and therefore only an option for rich countries. Seawalls were not high enough to protect communities devastated by the Fukushima tsunami in 2011.
	Prevent building in low-lying coastal areas exposed to tsunamis.	Major practical and economic difficulties, including existing fixed capital investments, inertia and resistance of local communities.

Chapter 2

Now test yourself

1 Insolation is (a) reflected from clouds and surfaces such as ice and sand, (b) absorbed by gas molecules in the atmosphere, e.g. ozone, (c) scattered by dust molecules and dust particles.

2 Solar radiation is short-wave, mainly in the visible spectrum; terrestrial radiation is long-wave, mainly infra-red.

3 Hadley cell:

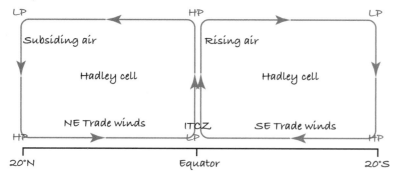

4 (1) The concentration of uplands and mountain ranges in the west, (2) proximity of the British Isles to the west and the North Atlantic Ocean.

5 An air mass is a large homogeneous body of air, with uniform temperature, humidity and lapse rate.

6 Air in depressions is converging and rising throughout the lower atmosphere. Air in anticyclones is subsiding towards the Earth's surface.

7 The savanna climate has a wet and dry season. The further from the equator, the shorter the wet season. The wet season corresponds to a 'summer' period of high sun: the dry season is the winter period of low sun.

8 (a) The Intertropical Convergence Zone (ITCZ) is an area of permanent low pressure within 10° of the equator where the trade winds converge. This creates the large-scale uplift of air that powers the general atmospheric circulation. (b) The subtropical high is a zone of permanent high pressure, situated between 20° and 30° latitude. In this zone, air subsides throughout the troposphere towards the surface, resulting in clear skies, low humidity and arid and semi-arid conditions.

9 Tropical cyclones are powerful storms that form over the ocean between latitudes 8° and 20°. They derive their energy from the warm surface waters of the ocean (at least 26°C to 27°C). They develop as tropical disturbances or waves which deepen and acquire a circular rotation (counter-clockwise in the Northern Hemisphere) due to the Earth's Coriolis effect.

10 Three hazards caused by tropical cyclones are: (a) hurricane-force winds, (b) torrential rain and river flooding, (c) storm surges.

11 Energy exchanges in urban areas differ from rural areas in the following ways: (1) more insolation is absorbed by the fabric of urban areas (brick, stone, tarmac), (2) less insolation reaches urban areas because of atmospheric pollution, (3) urban areas release heat energy through combustion.

12 (a) Urban heat islands describe higher temperatures (particularly at night) in still air conditions in urban areas compared with the surrounding countryside. (b) Photochemical smog develops in large urban areas in sunny weather and anticyclonic conditions. It restricts visibility and creates health risks to people suffering respiratory illnesses. It is caused by a chemical reaction between sunlight and air pollutants, particularly nitrogen dioxide from motor vehicles exhaust fumes.

13 Evidence of global climate change in the past 20,000 years is found in: (1) sea-floor sediments, (2) ancient pollen, (3) tree rings.

14 The greenhouse effect describes the response of the atmosphere to incoming solar radiation and outgoing terrestrial radiation. While most incoming, short-wave solar radiation is not absorbed by the atmosphere a significant proportion of outgoing, long-wave terrestrial radiation is (by carbon dioxide, methane, water vapour and other greenhouse gases). Some of this absorbed energy is re-radiated to the Earth and warms the atmosphere. Rising inputs of carbon dioxide due to human activity in the past 200 years have added to this absorption and warming, creating an enhanced greenhouse effect.

15 Global warming is a transboundary issue because, regardless of the source of greenhouse gases responsible for global warming, the impact is worldwide, affecting all countries.

16 Global warming could result in (a) worldwide increases in sea level, (b) loss of biodiversity, (c) shortages of water because of climate change and shrinking glaciers, (d) reduced agricultural food production and food shortages, (e) spread of tropical diseases, (f) changes in surface ocean currents.

17 (a) The Kyoto Protocol is the first international agreement to try to limit carbon emissions. Countries that signed the protocol have targets to reduce emissions by 2012. The protocol will not be renewed after 2012 and no agreement has been reached on a new treaty to replace Kyoto. (b) Carbon trading is an international scheme to reduce carbon emissions. Businesses are allocated carbon credits that limit their carbon emissions. If they exceed their limits they can buy carbon credits from businesses whose emissions are lower than their allocation. An international market in carbon trading has emerged in the past 10 years, based in London.

Check your understanding

1 Latitudes 0° to c. 40° (Earth and atmosphere) have an energy budget surplus (i.e. they have an excess of incoming solar radiation over outgoing terrestrial radiation). Those latitudes polewards of 40° have a negative energy budget over the year. The result is the poleward transfer of surplus heat energy from low latitudes to middle and high latitudes. This global imbalance of energy drives the general atmospheric circulation comprising wind systems and storms (e.g. tropical cyclones, depressions).

2 The British Isles, situated on the extreme northwestern fringes of Eurasia, at the boundary between the ocean and the continent, is exposed to air masses from all directions. Many air masses are oceanic in origin; they are humid, cloudy and rain-bearing, and associated with moderate temperatures. However, on approximately one day in four, continental air masses invade the British Isles. They bring settled and drier conditions, but more extreme temperatures.

3 Cross-section through a depression:

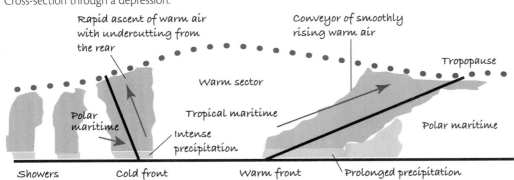

4 Global warming could result in significant decreases in rainfall, greater uncertainty of rainfall, longer periods of drought. Higher temperatures will increase water loss through evaporation. Marginal rain-fed farmland could be abandoned. Overexploitation of diminishing water, soil and vegetation resources could lead to widespread desertification. Crop yields will fall and food insecurity increase. There will be large-scale migration from rural areas to towns and cities. There will be a loss of biodiversity, with many large mammalian species facing extinction. Increasing CO_2 levels will favour woodland at the expense of grass, putting further pressure on mammalian herbivores.

5

Hazard	Description
Hurricane-force winds	Category 5 tropical cyclones have winds in excess of 119 km h^{-1}. High winds cause structural damage to property, bring down power cables, trees, etc.
Storm surges	Storm surges up to 10 m above normal sea level, driven by strong winds and low pressure, flood low-lying settlements and farmland
Torrential rain	Torrential rain causes rapid runoff and river flooding
Mass movements	Heavy rain saturates slopes, causing slope instability with resulting landslides, mudslides, mudflows, earthflows, etc.

6 Human activity creates an enhanced greenhouse effect by increasing levels of carbon dioxide, methane and other greenhouse gases (GHGs) in the atmosphere. These gases absorb long-wave radiation (heat) emitted from the Earth. Some of this heat is re-radiated back to Earth, raising temperatures. Increases in GHGs caused by human activity are due to: (1) burning fossil fuels — coal, oil, natural gas, (2) deforestation — less carbon stored in trees and more released to the atmosphere, (3) intensive agriculture — particularly livestock farming — emitting methane, (4) accelerated decay of peat and other plant remains as rising temperatures melt permafrost in the Arctic and sub-Arctic.

Chapter 3

Now test yourself

1 Ecosystems are holistic in the sense that their components — animals, plants and the physical environment — are highly interconnected. Because of this interconnectivity, ecosystems behave as an integrated whole. Changes in one part of the system have knock-on effects that lead to change in many other areas.

2 Simple food chain

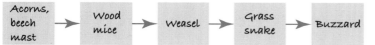

3 The main stores of nutrients are rocks, soil, dead organic matter, living organisms, atmosphere.

4 Primary plant succession occurs on new, previously unvegetated surfaces such as sand dunes, mudflats, lava flows, etc. Secondary succession occurs where the primary vegetation cover has been destroyed, e.g. by fire.

5 Climatic climax is the final stage of succession, with vegetation controlled by, and in equilibrium with, climate. In subclimax and plagioclimax vegetation forms, other environmental and human factors determine vegetation characteristics. Subclimax is the final stage of succession where non-climatic factors such as slope, soil or drainage determine the ultimate character of the vegetation. Plagioclimax is the final equilibrium stage where human activity is the dominant influence on vegetation type (e.g. heathland, moorland).

6 The problems for pioneer species such as glasswort and cord grass on mudflats include: (1) inundation for several hours at high tide, (2) high levels of salinity.

7 Heather moorland differs from natural ecosystems because it has (1) lower net primary productivity, (2) less biodiversity, (3) shorter food webs, (4) greater instability.

8 Equatorial rainforest is found within 10° of the equator. The largest areas of rainforest are the Amazon basin, the Congo basin and Indo-Malaysia.

9 (1) High levels of cloudiness in the rainforest. (2) Combination of overhead sun for up to two months a year and clear skies (deserts have clear skies for most of the year) produce higher temperatures in hot deserts.

10 Temperatures show little seasonal variation in the rainforest because (1) the sun is always at a high angle at noon (66–90°), (2) daylight hours are almost constant throughout the year, (3) high humidity and cloud cover help retain heat (absorb and reflect outgoing terrestrial radiation).

11 In the rainforest the soil has only a small nutrient store. Most nutrients are stored in the stems of forest trees. The forest survives by an efficient and rapid recycling of nutrients contained in dead organic matter (leaves, branches, stems). These nutrients, released by rapid decomposition in the warm, humid climate, are quickly removed from the soil by the shallow feeding roots of the forest trees.

12 Indigenous people exploit the rainforest sustainably by practising shifting cultivation, hunting wild animals, gathering fruits and roots. Pressure on the rainforest by indigenous groups is small because of their low densities and simple technologies. As a result indigenous economies are sustainable.

13 (1) Agriculture, including agribusiness growing soya and other crops, commercial ranching and peasant farming. (2) Mining, often by open-casting which completely destroys the rainforest and pollutes streams and rivers. (3) Large-scale logging for valuable hardwood timbers such as mahogany and ebony. (4) Road building to open up rainforest areas for economic development.

14 Urban areas provide opportunities for wildlife such as new habitats — gardens, parks, nature reserves, ponds, high-rise buildings for nesting, wasteland colonised by wild plants, insects, mammals and birds; food sources — garden bird feeders, scavenging foxes, feral pigeons predated by falcons.

15 Alien species such as Himalayan balsam, Japanese knotweed, mink and cane toads (in Australia) evolved in other ecosystems where their numbers were checked by natural predators. Introduced into other ecosystems and lacking natural checks, their populations grow rapidly. As a result they often out-compete native species, upsetting the natural balance, causing widespread disruption and disequilibrium.

16 In many ecosystems biodiversity and sustainability are under threat because of destruction of habitats by human activity, global warming and climate change, pollution, the introduction of alien species.

17 Fragile environments are easily degraded by human activity because of low biodiversity; low plant density, with incomplete vegetation cover; extreme weather events; extreme climatic conditions which slow plant growth and regeneration.

18 (1) Reduce the pressures of resource exploitation, e.g. fence-off open rangeland, reduce livestock densities. (2) Control population growth of indigenous people (family planning). (3) Introduce new technologies, e.g. zero cultivation, kerosene stoves to replace firewood. (4) Designate conservation areas where environmental protection has priority. (5) Only allow development that fulfils government criteria for sustainability.

Check your understanding

1 Energy declines with distance from the site of primary production in food webs or chains because 90% of energy is lost to respiration (and some to predation) at each trophic level. This loss of energy is reflected by a reduction in population numbers and biomass levels along food chains. It explains why populations of top predators, at the end of food chains, are relatively small.

2 The pioneer colonisers on mudflats are algae and plant species such as glasswort. These species encourage sedimentation, slowing the movement of tidal currents containing large amounts of suspended sediment. As the surface of mudflats rises, the period of inundation on the tidal cycle and salinity levels start to fall. This allows other species such as cord grass to invade. Further sedimentation ameliorates conditions until salt marsh, only inundated on spring tides, and colonised by plants such as sea aster, sea blight and sea plantain, develops. Salt marsh has complete vegetation cover, low levels of salinity, and shallow soils.

3 The temperate deciduous forest in vertical section (structure) comprises a canopy of mature trees (e.g. oak, ash, elm), an understorey of smaller trees such as birch, holly and hawthorn, and a ground vegetation that includes such species as brambles, bluebells, ransomes, etc. This structure is determined by light intensity, which in summer decreases towards the forest floor. Because the forest is deciduous, light intensity on the forest floor is highest in the spring. Most forest herbs flower before the canopy closes towards the end of May.

4 Biodiversity in the equatorial rainforest is greater than in any other terrestrial ecosystem because: (1) high temperatures and abundant rainfall throughout the year impose few climatic limits on plant growth. Plants do not need to evolve specialist structures to survive. (2) Food for insects and other animals is available all year round. (3) The rainforest is an ancient ecosystem that has survived unaltered for millions of years; it has provided a stable environment which has encouraged evolution and biodiversity.

5 Shifting cultivation mimics the rainforest ecosystems because: (a) it is a polyculture — several crops are grown simultaneously on cleared plots, giving stability; (b) it imposes limited pressure on the environment and is therefore sustainable; (c) the diversity of crops grown makes shifting cultivation, like the rainforest, highly productive.

6 Unsustainable development includes massive deforestation due to commercial logging, commercial farming, mining, etc. This type of development results in deforestation and either the outright destruction or degradation of the rainforest ecosystem. More sustainable use of the rainforests is achieved through conservation and ecotourism, e.g. in Costa Rica and elsewhere. Sustainability might also be achieved by giving farmers and governments economic incentives to conserve rainforests (subsidies, carbon credits, debt forgiveness, etc.).

Chapter 4

Now test yourself

1 (a) A mega city is a large city with a population of more than 10 million people, e.g. Buenos Aires. Its functional status is national or regional, but not global. (b) A world city is also a large city but supports a range of functions that play a vital role in the global economy, such as international finance and other producer services. The leading world cities are New York, London and Tokyo.

2 Globalisation encourages the growth of large towns and cities by: (1) liberalising and increasing international trade, which is articulated through major urban centres; (2) stimulating rural–urban migration, urbanisation and international population movements; (3) promoting industrialisation (e.g. China) through easier access to world markets.

3 Urban growth is simply an increase in the urban population of a country or region. Urbanisation is an increase in the proportion of urban dwellers. In the absence of migration, if a rural population increases faster than an urban population, urbanisation will not occur (i.e. the proportion of urban dwellers will fall).

4 Rapid urban growth is LEDCs is due to a combination of (a) natural increase, (b) rural–urban migration.

5 A greenfield site is land not previously used for urban development; a brownfield site is land that has previously been used by industry or commerce or housing, and has been redeveloped/reclaimed. Most brownfield sites are in urban areas. Greenfield sites are usually in rural or semi-rural locations (e.g. rural–urban fringe).

6 Urbanisation is a shift in population distribution from rural to urban areas (i.e. an increase in the proportion of urban dwellers in a country, region or local area). Counterurbanisation is a population shift in the opposite direction (from urban to rural), resulting in an increase in the proportion of rural dwellers.

7 Deindustrialisation is an absolute decline of industry in a country or region. Factories, mines, etc. are closed (most often because of international competition) creating high unemployment and industrial dereliction. Massive deindustrialisation hit UK regions such as central Scotland, northeast England and south Wales in the 1980s.

8 Deindustrialisation has the greatest impact in areas of economic specialisation. Cities (and regions) with excessive dependence on a narrow range of industries suffer massive economic problems (e.g. unemployment) when these industries collapse. More diversified economies are less affected by the structural decline of industry.

9 Property-led urban regeneration describes investment in new factory buildings, offices, business parks and essential infrastructure (most often using public money) to attract private investment and kick-start economic growth.

10 Urban decentralisation has increased inequality because: (1) new retailing formats have been concentrated in the suburbs, e.g. retail parks, supermarkets; (2) many businesses (e.g. offices) have located in suburban areas. These changes have tended to reduce access to retailing and job opportunities for inner city residents.

11 Product retailing is less important in city centres than it was 20–25 years ago. Leisure and entertainment services such as restaurants, cafés, clubs, etc. have replaced many product retailers. High streets have been refurbished (e.g. Birmingham) and made more pedestrian-friendly; in larger city centres there has been investment in covered shopping malls. There has been a movement back to central areas for residence (especially by singles and younger adults), with new multistorey apartments (often in waterfront locations) on the fringes of the CBD.

12 In the UK many local authorities are running out of space for landfill sites; landfill may contaminate the environment, with leakage of pollutants (in drainage or in the atmosphere); landfill sites may occupy land that could be used for other purposes (recreation, agriculture, forestry).

13 The benefits of recycling urban waste are less waste transferred to landfill, increased sustainability and less demand for new resources (e.g. paper, metal), reduced carbon emissions.

14 Traffic congestion imposes costs on the environment, society and the economy. Environmental costs — increased emissions of nitrogen oxides, carbon monoxide, particulates; social costs — increased risks to human health caused by pollutants, especially particulates, noise pollution; economic costs — increased fuel consumption, cost to businesses and individuals of time wasted in traffic queues.

15 Build more roads (ring roads, motorways) to speed flows of traffic; increase investment in public transport (rail, trams, buses); restrict movement of vehicles into city centres by imposing congestion charges or offering alternatives to private motorists (e.g. park-and-ride).

Check your understanding

1 World cities are needed to control and manage the global economy. Decisions that affect the global economy are made in world cities (e.g. investment, finance), often by TNCs which are headquartered there. As globalisation has extended (because of better telecommunications, transport, trade liberalisation) the role of world cities in the global economy has increased.

2 The main drivers of urbanisation in LEDCs are poverty and lack of opportunity in rural areas; industrialisation, linked to globalisation (e.g. China); declining mortality and rapid natural population growth; foreign direct investment, often concentrated in the largest cities.

3 The main changes in the internal distribution of population in cities in MEDCs: (1) decentralisation — movement of higher-income groups from inner city areas to the suburbs and commuter belt; (2) gentrification of some inner city districts by higher-income groups; (3) increase in residential properties (e.g. apartments — either new build or conversions of old industrial buildings) to house young people and singles who desire to live close to work and services in city centres; (4) movement of suburban residents to commuter villages in the exurbs and rural–urban fringe.

4 Some factors responsible for recent urban decline in MEDCs include (1) social dysfunction of some neighbourhoods — high crime levels, poor schools; (2) traffic congestion and air pollution in inner urban areas; (3) decentralisation of employment and services, sometimes to locations in the rural–urban fringe or on the edge of cities; (4) deindustrialisation and the decline of staple industries (e.g. Detroit).

5 Problems of urban decline in MEDCs have been tackled by urban regeneration schemes. These schemes often target the most deprived and run-down areas of cities (areas of multiple deprivation and environmental dereliction). Property-based schemes were

popular in the UK in the 1980s (Urban Development Corporations, Enterprise Zones). A more recent example is the New East Manchester scheme. Regeneration of this type is usually based on a partnership of public and private capital.

6 Planners in the UK have responded to the relative decline of retailing in city centres by: major new city centre investments, e.g. Bull Ring in Birmingham, Trinity Centre in Leeds; pedestrianisation; rebranding CBDs as 24 hour service centres; imposing planning restrictions on new out-of-centre retail developments that could potentially take trade from the CBD.

7 Cities are becoming more environmentally sustainable by: promoting recycling at the expense of landfill; encouraging residents to use public transport (e.g. park-and-ride) and bicycles (cycle lanes, bicycle hire); reducing carbon emissions (e.g. better insulation in homes, installing solar panels); redeveloping brownfield sites; creating green corridors for recreation and wildlife.

Chapter 5

Now test yourself

1 Quality of life is influenced by income, housing, access to services, access to healthcare, physical environment, leisure time, etc.

2 These terms suggest that a country's economic status is (a) not fixed and (b) part of an extended scale of development; that a country may be in a transitional stage of development before proceeding to the next level. Thus 'high income' countries implies the existence of 'middle' and 'low income' countries. 'Emerging economies' suggest that such countries were formerly in a pre-development stage and will eventually achieve full 'emergence' or development. 'Newly industrialising' suggests a continuum of development with countries at pre-industrial, industrial and post-industrial stages.

3 Development and demographic change — population growth rates fall as fertility decline matches the decline in mortality; population ages. Development and social change — smaller families, improved gender equality, advances in education levels and healthcare.

4 Globalisation is driven by increased volumes of world trade, a result partly of trade liberalisation; creation of trade blocs such as the EU; competition in manufacturing from emerging economies/NICs; advances in IT and telecommunications; the economic power of TNCs, etc.

5 Favourable demographics (young workforces, larger domestic markets), natural resources (especially Brazil, Russia and South Africa), high levels of foreign inward investment.

6 There is extreme poverty in sub-Saharan Africa because it is historically isolated, historically exploited by colonialism, has poor governance and political instability, many landlocked states, low levels of foreign direct investment, tropical diseases and AIDS.

7 Social problems in the world's poorest countries:

Health	Many of the world's poorest countries suffer high levels of HIV/AIDS prevalence. Malaria and TB are major causes of mortality and morbidity. Inadequate healthcare provision (both primary and secondary). High morbidity limits economic output
Nutrition	Poverty results in unbalanced diets, deficient in proteins and fats. Malnutrition is widespread. Food shortages, which in extreme conditions cause famine, create undernutrition. Poor nutrition weakens immune systems and results in chronic illness
Education	In some countries children have, on average, only 2 or 3 years of primary education. Adult illiteracy is widespread
Gender equality	Girls often more poorly educated than boys. Women often have little control over family size in male-dominated societies, little power, and fewer opportunities for employment

8 Human capital describes the skills of a workforce, which are vital for economic success. Most important are educational skills. High levels of skill, labour flexibility and motivation are major attractions for potential investors.

9 Many poor countries fail to develop human capital because of inadequate educational systems, discrimination against women, poor health and healthcare systems, political unrest and civil war.

10 The advantages of trade blocs such as the EU to member states include free trade between member states; movement of people/ labour force between member states; common policies for the environment, agriculture, etc.; financial assistance to less favoured regions.

11 TNCs are global corporations that have production functions (factories, offices) in several countries and operate in global markets (e.g. Coca-Cola, Ford, BP, HSBC, Microsoft). TNCs are large-scale organisations, with massive turnovers and thousands of employees. They are major players in the global economy, are responsible for a large proportion of international trade, and have considerable political influence as well as economic power.

12 TNCs have influenced globalisation by investing in production functions worldwide, targeting global markets, accounting for a large volume of international trade both within and between TNCs.

13 LEDCs often face obstacles to expanding their international trade because of: external tariff barriers that restrict entry to national and supranational markets; unfair competition — price subsidies on foreign imports handicap the growth of domestic producers; poorly developed transport infrastructure (to facilitate exports); reliance on a narrow range of primary products (e.g. metallic ores,

unprocessed food) whose value has failed to keep pace with prices for manufactured goods; huge price fluctuations for primary products on world markets.

14 Bilateral aid is aid given by a single donor country to a recipient country. Multilateral aid is aid donated by an agency such as the UN and the World Bank to a recipient country.

15 International aid has provided limited benefits for the poorest people in recipient countries because aid has often been directed at large-scale, capital-intensive projects (e.g. dams); corruption and poor governance has squandered aid money; many aid projects, while aimed at the poorest groups, have not been sustainable.

16 Ecotourism minimises its impact on natural environments and ecosystems and creates economic opportunities for local people. Sustainable tourism should, in theory, have no adverse impact on the local environment (e.g. degraded ecosystems, water resources, erosion) and local cultures. Mass tourism caters for very large numbers of tourists; developments often degrade the local environment and local cultures; its priorities are economic (e.g. employment, foreign exchange revenue) rather than environmental or cultural.

Check your understanding

1 Countries are in many different stages of economic development. It is simplistic to view countries as either 'developed' or 'less developed'. The World Bank recognises four categories of development — high income, upper middle income, lower middle income and low income. Even this classification depends on arbitrary boundaries. Defining types of countries based on their development characteristics is a convenient way of understanding the complexity of development. It is, however, a gross generalisation — the reality is that there is a continuum of development, not a series of discrete stages.

2 (1) Fewer obstacles to trade — removal of tariffs and subsidies (WTO); (2) growth of TNCs targeting global rather than national markets — economies of scale; (3) developments in transport (massive container ships) and telecommunications; (4) the huge economic growth in China, India and other NICs/emerging economies.

3 TNCs are the most powerful force in the globalised economy. A large proportion of world trade takes place within and between TNCs. TNCs are the main instruments of foreign direct investment, which creates jobs, increases consumer spending and generates new markets. TNCs such as McDonald's, Coca-Cola and Microsoft change cultural preferences and help to create homogeneous consumption and markets.

4 The main obstacles to development in the world's poorest countries include poor governance, geographical isolation, poor social and transport infrastructure, poor levels of education and healthcare, incidence of disease, uncompetitive industries and other types of production, poor governance, corruption and political instability.

5 The development of free trade should open markets to exporters from LEDCs. Ending export subsidies by MEDCs would give economic activity in the developing world a more competitive edge. Fair trade should give poor farmers and other producers in LEDCs a better return on their exports and raise standards of living. International aid is most effective when it targets poor communities directly. Small-scale projects such as building rural schools, family planning clinics, improving irrigation techniques, providing clean water supplies, etc. can improve the lives of the poorest communities. In the past large-scale projects have done little to help the poor. Educated and more healthy people are more productive and more likely to achieve their potential.

6 Sustainable tourism requires good governance, with long-term planning (e.g. designating conservation areas, protecting local cultures, giving local people a stake in sustainable tourism). Governments must accept that sustainable tourism is only possible on a limited scale and that economic returns are generated by a relatively small number of visitors with high per capita spending.

Chapter 6

Now test yourself

1 Localism describes local issues and conflicts relating to villages or small towns that affect small communities. Regionalism concerns issues that relate to a larger geographical area and involves a common sense of history, geography and culture rather than just place of residence (as in localism). Nationalism involves issues that relate to large groups with shared language, history, ethnicity or race.

2 A nation is a large group of people who share a common language, history, heritage, ethnicity or race (e.g. the Basques in northern Spain). A nation state is where a national group has its own government and sovereignty.

3 Ideology is a theory or set of beliefs or principles on which a political system is based. Examples include democracy and communism.

4 Conflicts can be resolved by negotiation, mediation, appeals/government enquiries, war, devolution of power, territorial partition, etc.

5 Separatism concerns nationalist groups within a country that campaign for their own representation, government or sovereignty. The aim may be devolution of power or full independence (e.g. Scottish National Party).

6 Conflict at Foveran has developed because a new golf course will destroy part of the natural environment and ecosystem in the Sands of Forvie dunes (which have protected status); local people regard the development as intrusive and likely to bring unwanted change; the Scottish government supports the development because of benefits it will bring to the local economy.

7 The Arab–Israeli conflict centres on competing claims for the same territory. Thousands of Palestinians were dispossessed with the creation of the Jewish state of Israel in 1948. The Jewish claim to Palestine dates back two millennia, when the territory was the

Jewish homeland. The Palestinians have a similar historic claim which is centuries old. Since 1948 there have been four wars between the Israelis and the surrounding Arab states that support the Palestinian cause. A negotiated settlement of the dispute has proved elusive.

8 Social issues — Palestinian refugees in neighbouring Arab states (Jordan, Lebanon, etc.); appalling overcrowding and insanitary conditions in Gaza. Economic issues — poverty in Gaza and among the Palestinians in general; unemployment. Environmental issues — water resources, and the division of water between Israel, Jordan and Syria; environmental impact of over-extraction of water in the Jordan River basin; loss of wildlife habitat in densely populated Israel and the West Bank.

9 Ethnic minorities are defined by their separate cultural identities — language, religion, customs, etc. Some ethnic minorities may belong to distinctive racial groups, though ethnicity is not defined by race.

10 (a) Multiculturalism is the idea that several different cultures can co-exist peacefully and equitably in a country (rather than one national culture). (b) A multicultural society is one that has a diverse ethnic structure. Ethnic diversity and differences are promoted by governments committed to policies of multiculturalism.

11 The UK's multicultural society developed from: government policies of multiculturalism, especially in the late twentieth century; large-scale immigration of diverse ethnic groups from the Caribbean, south Asia, Africa, the Middle East and eastern Europe; globalisation and the expansion of the EU; traditions of tolerance, providing refuge for people fleeing political persecution; close political ties arising from Empire and Commonwealth connections.

12 Uneven geographical distributions of ethnic groups in the UK at the regional scale are influence by the availability of employment and the existing location of ethnic groups. At an intra-urban scale, location is influenced by availability of affordable housing, proximity to ethnic services (shops, mosques, temples, etc.), location of similar ethnic groups, security and defence against perceived threats from the host society.

13 Poverty at the global scale is concentrated in the developing world, most obviously in sub-Saharan Africa. Poverty is also widespread in south Asia and Central America.

14 Global distribution of poverty:

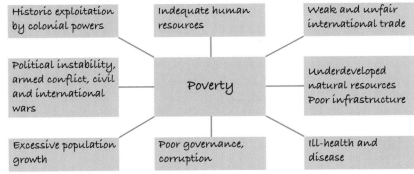

15 The Millennium Development Goals, outlined in 2000 and set by the UN, established targets on global poverty, hunger, gender equality, education, environment and health to be met by 2015.

16 The Millennium Development Goals are unlikely to be achieved by 2015. Possible reasons include the global economic recession (2008–2011), political instability in many parts of the developing world, continued rapid population growth in the poorest countries. Several goals were unrealistic and were never likely to be achieved by 2015.

17 Lack of national security deters foreign direct investment and international aid, vital to give economic development momentum. At the same time, poverty, poor governance and lack of development provide an environment in which political dissent, separatism and terrorism can flourish. Political chaos and anarchy in countries such as Somalia and DR Congo flourish against a background of poverty and lack of development.

Check your understanding

1 Palestinians and Israelis have conflicting territorial claims to Israel. Both have occupied the territory in the past — the Jews 2,000 years ago; the Palestinians until the twentieth century. Diplomatic solutions centre on territory, i.e. partition, with Israel reverting to its pre-1967 borders and the 'occupied territories' forming a new Palestinian state. There are also unresolved social issues — Palestinian refugees in surrounding Arab states, inequalities between Jews and Arabs in Israel (e.g. Gaza and West Bank). The importance of water resources in the conflict is less obvious. Israel is water-deficient and the Jordan Basin and Sea of Galilee provide important water resources. If Israel cedes the West Bank it would lose full control of these resources.

2 Many sovereign states were created in the twentieth century from former colonies (especially in Africa). Their borders were often drawn for political convenience, with little concession to national and tribal loyalties. Several other long-established states resulted from centuries of war, conquest or colonisation, e.g. minority nations such as Wales, Scotland and Ireland in the UK, colonisation of North America and the Indian nations.

3 Minority national groups, particularly those forced into political union by conquest, often occupy relatively isolated and marginal territories, e.g. the Basques, the Scots. Separatist movements fighting guerrilla wars often base their operations in strategically inaccessible locations such as mountains and rainforests (e.g. Taliban in Pakistan). However, with modern telecommunications separatist movements can also operate effectively (and covertly) in large urban centres throughout a country.

4 Environmental issues at a local scale arise from developments that cause loss of habitat, threats to wildlife, loss of countryside, pollution, etc. Local objectors are often motivated by self-interest or nimbyism. Proponents of development argue that degrading effects on the environment are offset by economic benefits to the local and wider community. Environmental issues are usually resolved by discussion and consultation between planners and other interested parties, with final decisions made by local representatives. In the UK, major issues that cannot be resolved in this way sometimes go to public enquiry.

5 Economic disadvantages — immigrants are likely to be young adults, many with children. This, together with higher fertility among some immigrant groups, makes heavy demands on education and healthcare services; jobs previously filled by native people may be taken by better qualified immigrants. Social disadvantages — where cultural differences between immigrants and the host population are large, prejudice, segregation and minimal integration are more likely. Isolation and lack of integration can disadvantage the children of immigrants educationally, who may struggle to find employment as young adults. The overall cohesiveness of society and a sense of national belonging might be weakened.

6 Segregation among ethnic minority groups is due to both positive and negative factors. Positive factors include a preference to live close to people of similar culture and language, access to ethnic services such as places of worship and shops, access to employment. Negative factors include reducing threat and prejudice (perceived or real) from the host society, poverty and the affordability of housing, discrimination in the private housing market.

Chapter 7

Now test yourself

1 Possible primary sources are fieldwork (observation, measurement, questionnaires), census documents, directories, etc. Possible secondary sources are articles, textbooks, internet, maps, etc.

2 A theory is a set of statements or principles that has been proved valid. Theories such as systems theory and central place theory are often used in geographical investigations. Geographical background might cover a detailed description of the study location and its surrounding area, and ideas and concepts that are relevant to the investigation (e.g. fluvial processes and Hjulström's curve, counterurbanisation, land use in the rural–urban fringe, etc.).

3 Risk assessment is needed before starting fieldwork. On completion the risk assessment should be vetted by a teacher or supervisor. Precautions designed to reduce risk often involve wearing suitable protective clothing, carrying a mobile phone, leaving details of the itinerary and contact numbers with a responsible adult, working in groups, carrying a first-aid kit, torch and emergency rations, avoiding (through risk assessment) working in hazardous areas.

4 The location chosen for an investigative study could be influenced by a range of factors — scale, the likelihood that it will yield the appropriate amount and type of data, accessibility, risk.

5 Fieldwork data are usually collected as samples. Samples must be as accurate and reliable as possible. This can be achieved by (1) ensuring that a sufficiently large sample of data is collected, (2) that sample data are collected objectively and that the sampling method chosen produces data that represent the population. Possible methods of statistical sampling are random sampling, systematic sampling, stratified sampling. If the data have a spatial dimension, samples may be collected at points, for areas or along lines (transects).

6 Criteria that can be used to justify choices of graphical and mapping methods are: visual impact, clarity and accuracy, degree of generalisation, ease of construction, appropriateness of the method to the type of data.

7 Analysis of primary data could rely on graphical and/or mapping techniques, and/or statistical analysis. Data presentation methods can make patterns or trends clearer. Statistical analysis will help to generalise data (central tendency measures) or determine the reliability of outcomes (i.e. significance) and will sometimes reveal hidden patterns in the data. The choice of graphical and mapping methods will use the criteria listed in (6) above. Statistical methods of analysis depend on the nature of the hypothesis: a hypothesis of 'difference' will use the U test or Chi-Squared test; a hypothesis of association will use correlation.

8 The conclusion to an investigation will include a list of the main findings. Findings will be in the form of statements, e.g. channel capacity increased with distance downstream. However, findings can be negative as well as positive, e.g. no relationship was found between channel width and channel gradient.

9 The introduction to a geographical investigation will set out the aims of the enquiry (usually a question or a hypothesis). Few geographical investigations are straightforward because many different factors influence outcomes (e.g. appropriateness of the hypothesis, amount of data collected, timing of data collection, representativeness of data, etc.) These problems give ample scope for discussion. Discussion should be critical, demonstrate awareness of the complexity of geographical phenomena and data collection through fieldwork.

Check your understanding

1 There are six main stages of geographical investigation: (1) devising a research question or hypothesis, (2) developing a plan or strategy, (3) data collection, (4) data description and presentation, (5) data analysis, (6) summary and conclusion.

2   The main methods of statistical sampling are random, systematic and stratified. Stratified samples are collected either randomly or systematically. If location is a key important attribute of the sample data, samples may be collected at points, in areas or along lines (transects). The sampling method chosen and its degree of usefulness will be influenced by the nature of the statistical population, the time and resources available for sampling, and the degree of accuracy required.

3   The relative value of dot maps and choropleth maps can be assessed against: their visual impact and their success in conveying spatial patterns and trends; accuracy in their representation of spatial distributions; degree of generalisation involved in their construction; difficulty of construction; ease and accuracy of statistical data extracted from them. Dot maps involve subjectivity in the placement of dots, the value given to each dot, and to the choice of dot size. Choropleth maps also have weaknesses. They include variability in the size and shape of areal units (e.g. administrative/census areas) which often affect a map's appearance, excessive generalisation of values in the largest areal units, and the artificial influence of the boundaries on the appearance and accuracy of the map. (i.e. abrupt changes at boundaries).

4   The arithmetic mean or average is the most accurate measure of central tendency. Its calculation uses all the values in a data set. However, the mean is not representative of data sets that contain extreme values (i.e. small values or large values). This is because the mean is influenced by the magnitude of each value in a data set. The median is the middle value in a data set arranged in rank order. It is more representative than the mean for skewed data, but its calculation uses only 50% of the values in the data set. The principal mode is the class in a frequency table or histogram with largest number of values. It is a more general measure than the mean or the median; being a class, it does not have a single value, and has limited use when data sets are bi-modal or tri-modal.

5   Statistical significance determines the probability that the outcome of statistical analysis (e.g. correlation or U test) has occurred purely by chance. In geography, if the probability of a result occurring by chance is less than 5%, we say it is statistically significant. Correlation is an inferential statistical test that is widely used in geographical investigation. It measures the association between two variables — an independent variable such as rainfall, that causes change in a dependent variable such as river discharge. The strength of an association or relationship is given by a correlation coefficient which varies from −1 (perfect inverse correlation) to +1 (perfect positive correlation). A correlation coefficient close to zero suggests little if any association. Because correlation analysis is usually based on sample data, the statistical significance of a correlation coefficient must be established by reference to tables of statistical significance.